Grade 3

Spelling
PRACTICE BOOK

Macmillan
McGraw-Hill

B

Macmillan McGraw-Hill

Published by Macmillan/McGraw-Hill, of McGraw-Hill Education, a division of The McGraw-Hill Companies, Inc., Two Penn Plaza, New York, New York 10121.

Printed in the United States of America

10 006 09 08

Contents

Unit 2 • Investigations

Unit 3 • Discoveries

Unit 4 • Determination

© Macmillan/McGraw-Hill

Unit 5 • Challenges

Unit 6 • Achievements

Name_____

Fold back the paper along the dotted line. Use the blanks to write each word as it is read aloud. When you finish the test, unfold the paper. Use the list at the right to correct any spelling mistakes.

1. _____
2. _____
3. _____
4. _____
5. _____
6. _____
7. _____
8. _____
9. _____
10. _____
11. _____
12. _____
13. _____
14. _____
15. _____

Review Words   16. _____

17. _____
18. _____

Challenge Words   19. _____

20. _____

1. clap
2. step
3. sick
4. rock
5. luck
6. crop
7. snack
8. mess
9. head
10. shut
11. miss
12. stamp
13. jump
14. click
15. pond
16. cat
17. man
18. can
19. bathtub
20. anthill

Name_____

Using the Word Study Steps

1. LOOK at the word.

2. SAY the word aloud.

3. STUDY the letters in the word.

4. WRITE the word.

5. CHECK the word.
   Did you spell the word right?
   If not, go back to step 1.

Find Rhyming Words

Circle the word in each row that rhymes with the word in dark type.

1. **sock**	truck	rock	sick
2. **dress**	mess	dust	mast
3. **trap**	track	clam	clap
4. **bump**	jump	junk	bunch
5. **fed**	hid	head	hide
6. **pick**	sick	sock	dock
7. **ramp**	fan	sand	stamp
8. **back**	snap	sneak	snack
9. **top**	tip	crib	crop
10. **brick**	clock	click	cluck
11. **kiss**	miss	mist	mask
12. **cut**	shut	cat	sat
13. **duck**	dark	luck	lark
14. **pep**	pop	step	stop
15. **bond**	plod	plop	pond

© Macmillan/McGraw-Hill

At Home: Review the Word Study Steps to help your child spell new words.

Name_____

clap	rock	snack	shut	jump
step	luck	mess	miss	click
sick	crop	head	stamp	pond

Write the spelling words that rhyme with the words below. Then circle the letter that spells the short vowel sound in each word.

1. block _____

2. tuck _____

3. bed _____

4. cut _____

5. flap _____

Vowel Power

Write the spelling words that contain each short vowel sound below.

short a

6. _____

7. _____

8. _____

short e

9. _____

10. _____

11. _____

short i

12. _____

13. _____

14. _____

short o

15. _____

16. _____

17. _____

short u

18. _____

19. _____

20. _____

clap	rock	snack	shut	jump
step	luck	mess	miss	click
sick	crop	head	stamp	pond

What's the Word?

Complete each sentence with a spelling word.

1. My sister makes a _____ when she packs for school.

2. Put a _____ on the letter before you send it.

3. I _____ my mom when I go to school.

4. After math class, it's time for a _____.

5. Ducks like to swim in a _____.

6. We _____ when the music ends.

7. I walk with my _____ held high.

8. Her shoes _____ on the floor when she walks.

9. The farmer grew a _____ of corn.

10. I don't like to _____ on cracks in the sidewalk.

Define It!

Write the spelling words that have the same meaning as the words below.

11. close _____

12. stone _____

13. ill _____

14. leap _____

15. chance _____

Name _____

Fold back the paper along the dotted line. Use the blanks to write each word as it is read aloud. When you finish the test, unfold the paper. Use the list at the right to correct any spelling mistakes.

1. _____
2. _____
3. _____
4. _____
5. _____
6. _____
7. _____
8. _____
9. _____
10. _____
11. _____
12. _____
13. _____
14. _____
15. _____

Review Words 16. _____

17. _____

18. _____

Challenge Words 19. _____

20. _____

1. date
2. fine
3. rose
4. lake
5. life
6. home
7. safe
8. rice
9. globe
10. plane
11. wise
12. smoke
13. grade
14. smile
15. come
16. clap
17. sick
18. crop
19. sneeze
20. escape

© Macmillan/McGraw-Hill

At Home: Help your child practice the words he or she missed to prepare for the Posttest.

Name_____

Using the Word Study Steps

1. LOOK at the word.
2. SAY the word aloud.
3. STUDY the letters in the word.
4. WRITE the word.

5. CHECK the word.
   Did you spell the word right?
   If not, go back to step 1.

Choose the spelling word that best completes the sentence.

1. My favorite flower is a _____.

2. We have to fly in a _____ to visit my grandparents.

3. I make sure to put the _____ on the top of my letters.

4. Dave swims in the _____ every summer.

5. I looked at a _____ to see the country where my pen pal was from.

6. Jill saw _____ at the top of the house and knew there was a fire.

7. My favorite meal for dinner is _____ and chicken.

8. My younger sister is in the first _____.

9. I asked my uncle to _____ over to help me with my homework.

10. To live a long _____ you should exercise and eat healthy food.

11. I wrote my brother a letter to ask him when he was coming _____.

12. My grandfather tells me to always do my homework so I can be _____ when I grow up.

13. I _____ every time I see my new puppy.

14. To be _____ I look both ways when I cross the street.

15. I felt _____ after a lot of rest.

At Home: Review the Word Study Steps to help your child spell new words.

Name_____

date	lake	safe	plane	grade
fine	life	rice	wise	smile
rose	home	globe	smoke	come

Write the spelling words that contain each long vowel sound below.

long *a*

1. _____
2. _____
3. _____
4. _____
5. _____

long *i*

6. _____
7. _____
8. _____
9. _____
10. _____

long *o*

11. _____
12. _____
13. _____
14. _____

Name_____

date	lake	safe	plane	grade
fine	life	rice	wise	smile
rose	home	globe	smoke	come

It Takes Three

Write a spelling word that goes with the other two words.

1. pond, sea, _____

2. world, Earth, _____

3. smart, clever, _____

4. tulip, daisy, _____

What Does It Mean?

Write a spelling word that matches each clue below.

5. The place where you live _____

6. The day of the year _____

7. Not a frown _____

8. Flying machine _____

9. Rises from a fire _____

10. Out of harm's way _____

11. A side dish _____

12. A class or year in school _____

13. Arrive _____

14. Feeling well _____

15. A person's time on Earth _____

Name _____

Fold back the paper along the dotted line. Use the blanks to write each word as it is read aloud. When you finish the test, unfold the paper. Use the list at the right to correct any spelling mistakes.

1. _____   1. fail

2. _____   2. bay

3. _____   3. pail

4. _____   4. ray

5. _____   5. plain

6. _____   6. tray

7. _____   7. trail

8. _____   8. May

9. _____   9. braid

10. _____   10. sway

11. _____   11. gray

12. _____   12. plays

13. _____   13. paint

14. _____   14. snail

15. _____   15. great

Review Words   16. _____   16. safe

17. _____   17. rice

18. _____   18. globe

Challenge Words   19. _____   19. lady

20. _____   20. afraid

At Home: Help your child practice the words he or she missed to prepare for the Posttest.

Whose Habitat Is It? • **Book I/Unit I**   13

Name _____

Using the Word Study Steps

1. LOOK at the word.

2. SAY the word aloud.

3. STUDY the letters in the word.

4. WRITE the word.

5. CHECK the word.
   Did you spell the word right?
   If not, go back to step 1.

Find and Circle

Where are the spelling words?

P	L	A	I	N	M	P	L	A	Y	S	O	E	H
A	B	R	A	I	D	A	G	R	E	A	T	S	F
I	A	T	S	K	C	I	R	A	Y	A	R	U	A
N	Y	S	N	A	I	L	A	N	P	M	A	Y	I
T	R	A	I	L	D	J	Y	S	W	A	Y	V	L

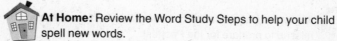

At Home: Review the Word Study Steps to help your child spell new words.

© Macmillan/McGraw-Hill

Name_____

fail	ray	trail	sway	paint
bay	plain	May	gray	snail
pail	tray	braid	plays	great

Write the spelling words that contain the matching spelling of the long *a* sound.

long *a* spelled *ai*

1. _____

2. _____

3. _____

4. _____

5. _____

6. _____

7. _____

long *a* spelled *ay*

9. _____

10. _____

11. _____

12. _____

13. _____

14. _____

15. _____

long *a* spelled *ea*

8. _____

Words Within Words

Add one letter to the word to create a spelling word.

16. _____ + ray = _____ or _____

17. _____ + nail = _____

18. _____ + way = _____

19. _____ + lays = _____

Name_____

fail	ray	trail	sway	paint
bay	plain	May	gray	snail
pail	tray	braid	plays	great

It Takes Three
Write a spelling word that goes with the other two words.

1. dish, plate, _____

2. pond, lake, _____

3. path, road, _____

4. June, July, _____

What Does It Mean?
Write a spelling word that matches each clue below.

5. Very good _____

6. Simple _____

7. A beam of light _____

8. Dull and gloomy _____

9. An animal that moves slowly _____

10. A bucket _____

11. What a kitten does with yarn _____

Past Tense
To form the past tense of a verb you usually add *-ed*. Put these words in the past tense:

12. braid _____

14. sway _____

13. paint _____

15. fail _____

Name _____

Fold back the paper along the dotted line. Use the blanks to write each word as it is read aloud. When you finish the test, unfold the paper. Use the list at the right to correct any spelling mistakes.

1. _____   **1.** gold

2. _____   **2.** bowl

3. _____   **3.** soak

4. _____   **4.** sold

5. _____   **5.** snow

6. _____   **6.** loaf

7. _____   **7.** roast

8. _____   **8.** coast

9. _____   **9.** scold

10. _____   **10.** coal

11. _____   **11.** slow

12. _____   **12.** grows

13. _____   **13.** show

14. _____   **14.** float

15. _____   **15.** blow

Review Words   16. _____   **16.** snail

17. _____   **17.** plain

18. _____   **18.** gray

Challenge Words   19. _____   **19.** window

20. _____   **20.** program

© Macmillan/McGraw-Hill

At Home: Help your child practice the words he or she missed to prepare for the Posttest.

Name_____

Using the Word Study Steps

1. LOOK at the word.

2. SAY the word aloud.

3. STUDY the letters in the word.

4. WRITE the word.

5. CHECK the word.
   Did you spell the word right?
   If not, go back to step 1.

Fill in the missing letters of each word to create a spelling word.

1. s h _____ _____

2. r _____ _____ s t

3. s c _____ l d

4. b l _____ _____

5. f l _____ _____ t

6. g _____ l d

7. g r _____ _____ s

8. c _____ _____ s t

9. s _____ l d

10. b _____ _____ l

11. c _____ _____ l

12. s l _____ _____

13. l _____ _____ f

14. s n _____ _____

15. s _____ _____ k

Choose the spelling word that best completes the sentence.

1. The boat will _____ on the lake.

2. _____ fell all night and covered the ground.

3. I like to have a _____ of ice cream for dessert.

At Home: Review the Word Study Steps to help your child spell new words.

© Macmillan/McGraw-Hill

Name_____

gold	sold	roast	coal	show
bowl	snow	coast	slow	float
soak	loaf	scold	grows	blow

Write the spelling words that contain the matching spelling of the long *o* sound.

long *o* spelled *ow*

1. _____
2. _____
3. _____
4. _____
5. _____
6. _____

long *o* spelled *oa*

10. _____
11. _____
12. _____
13. _____
14. _____
15. _____

long *o* spelled *o*

7. _____
8. _____
9. _____

Words Within Words

Write the spelling words that contain the small word.

16. old _____
17. oak _____
18. cold _____
19. oat _____
20. low _____

Name_____

gold	sold	roast	coal	show
bowl	snow	coast	slow	float
soak	loaf	scold	grows	blow

Words in Sentences

Write a spelling word to complete each sentence.

1. I had a _____ of soup for lunch.

2. We bought a _____ of bread at the store.

3. A penguin chick hatches and _____ up.

4. Pieces of ice _____ on top of the water.

5. Mark had to _____ his dog for digging up the flowers.

6. The ground in Antartica is covered in _____.

7. They used to heat houses with _____.

8. I brought my cat to school for _____ and tell.

9. There were many _____ necklaces in the window of the store.

10. The girls _____ cookies outside the store.

11. On her birthday, Maggie will _____ out the candles on her cake.

12. We had to _____ the sponges in water.

Opposite

Write the spelling word that is the opposite in meaning to the word below.

13. fast _____

14. hide _____

15. sink _____

16. praise _____

Name_____

There are seven spelling mistakes in this postcard. Circle the misspelled words. Write the words correctly on the lines below.

Dear Paula,

I told you I would send you a postcard! You sure do need to come here next summer. We had so much fun on our family vacation. We would sit on beaches that had sand the color of solid golde. The other penguins and I would play on the beach, flowte in the water, and soke up the sun. At night we would stay up late to listen to the singing of the whales while eating a big boal of ice cream.

I am sad that we have to leave the coste in a few days and go home to Antarctica. I hope there isn't too much sno at home. I will shoe you pictures when I get home.

See you soon,
Peter

1. _____   5. _____

2. _____   6. _____

3. _____   7. _____

4. _____

Writing Activity

Write about what you like to do on a cold or snowy day. Use at least three spelling words in your description.

_____

_____

_____

_____

_____

Name_____

Look at the words in each set below. One word in each set is spelled correctly. Look at Sample A. The letter next to the correctly spelled word in Sample A has been shaded in. Do Sample B yourself. Shade the letter of the word that is spelled correctly. When you are sure you know what to do, go on with the rest of the page.

Sample A:

Ⓐ roa
Ⓑ row
Ⓒ roaw
Ⓓ rowe

Sample B:

Ⓔ rowd
Ⓕ rowde
Ⓖ roade
Ⓗ road

1. Ⓐ gowld
   Ⓑ golde
   Ⓒ gold
   Ⓓ goald

2. Ⓔ bole
   Ⓕ boal
   Ⓖ boale
   Ⓗ bowl

3. Ⓐ soke
   Ⓑ sowk
   Ⓒ soak
   Ⓓ sok

4. Ⓔ sowld
   Ⓕ sold
   Ⓖ soald
   Ⓗ solde

5. Ⓐ snow
   Ⓑ snowe
   Ⓒ snoe
   Ⓓ snoa

6. Ⓔ lofe
   Ⓕ loaf
   Ⓖ loafe
   Ⓗ lowf

7. Ⓐ roast
   Ⓑ roste
   Ⓒ rowst
   Ⓓ rost

8. Ⓔ cowst
   Ⓕ coast
   Ⓖ coste
   Ⓗ coaste

9. Ⓐ scold
   Ⓑ scowld
   Ⓒ scoald
   Ⓓ scolde

10. Ⓔ coale
    Ⓕ caol
    Ⓖ cole
    Ⓗ coal

11. Ⓐ slow
    Ⓑ sloe
    Ⓒ slowe
    Ⓓ sloaw

12. Ⓔ grows
    Ⓕ groaws
    Ⓖ groze
    Ⓗ groaz

13. Ⓐ show
    Ⓑ shoew
    Ⓒ shoaw
    Ⓓ showe

14. Ⓔ floet
    Ⓕ flote
    Ⓖ float
    Ⓗ flowt

15. Ⓐ bloa
    Ⓑ bloae
    Ⓒ blowe
    Ⓓ blow

© Macmillan/McGraw-Hill

Name _____

Fold back the paper along the dotted line. Use the blanks to write each word as it is read aloud. When you finish the test, unfold the paper. Use the list at the right to correct any spelling mistakes.

1. _____
2. _____
3. _____
4. _____
5. _____
6. _____
7. _____
8. _____
9. _____
10. _____
11. _____
12. _____
13. _____
14. _____
15. _____

Review Words 16. _____

17. _____

18. _____

Challenge Words 19. _____

20. _____

1. mild
2. sky
3. pie
4. might
5. find
6. fight
7. ties
8. right
9. fry
10. tight
11. child
12. flight
13. bright
14. buy
15. dye
16. soak
17. bowl
18. gold
19. wind
20. children

At Home: Help your child practice the words he or she missed to prepare for the Posttest.

The Perfect Pet • **Book I/Unit I**   25

© Macmillan/McGraw-Hill

Name_____

Using the Word Study Steps

1. LOOK at the word.

2. SAY the word aloud.

3. STUDY the letters in the word.

4. WRITE the word.

5. CHECK the word.
   Did you spell the word right?
   If not, go back to step 1.

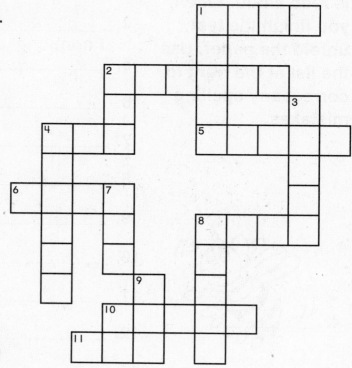

Crossword Puzzle

**Read the clues. Then use the
spelling words to complete the
sentences.**

ACROSS

1. The salsa was _____.

2. The plane _____ was long.

4. I ____ milk for my new kitten.

5. The pants were too _____.

6. David _____ his shoes.

8. I cannot _____ my winter coat.

10. My answer was _____.

11. She uses ____ to color her hair.

DOWN

1. He closed the door with all his _____.

2. Let's _____ fish for dinner.

3. The _____ missed his mother.

4. The sun is _____.

7. The bird flew up high in the _____.

8. The two angry dogs got into a _____.

9. We ate apple _____.

At Home: Review the Word Study Steps to help your child
spell new words.

Name_____

mild	might	ties	tight	bright
sky	find	right	child	buy
pie	fight	fry	flight	dye

Write the spelling words that contain the matching spelling of the long *i* sound.

long *i* spelled *ie*

1. _____

2. _____

long *i* spelled *igh*

3. _____

4. _____

5. _____

6. _____

7. _____

8. _____

long *i* spelled *y*

9. _____

10. _____

long *i* spelled *i*

11. _____

12. _____

13. _____

Name_____

mild	might	ties	tight	bright
sky	find	right	child	buy
pie	fight	fry	flight	dye

Analogies

Use the clues to find the correct spelling word. Write your answer on the line.

1. not spicy _____

2. to discover something _____

3. the opposite of *wrong* _____

4. to get something at the store _____

5. a young person _____

In the Dictionary

Many dictionary entries have sample sentences that show how the word can be used.

Complete each sample sentence with a spelling word.

6. The cook baked a peach _____.

7. The car's lights were very _____.

8. Birds fly in the _____.

9. I _____ come to the party late.

10. You use _____ to change the color of cloth.

Name_____

There are seven spelling mistakes in these paragraphs. Circle the misspelled words. Write the words correctly on the lines below.

"Mom, can I leave yet?" yelled Lisa. Tomorrow was Lisa's birthday and she was getting a new puppy. She loved to go to the pet store and look at the puppy she had picked out.

"Have you finished your dinner and had a piece of pye?" asked Lisa's mother.

"Yes mom," replied Lisa.

"Alright, you can go but you myght not be able to see your puppy because the store is closed," said Lisa's mother.

Lisa didn't even hear her mother because she ran rieght out the door to see her puppy.

Lisa skipped down the street. She thought about her new puppy and how happy she was that she could fynde a puppy she liked. When she got to the store she peeked in the window. There he was playing in his wire cage. The store was closed and the lights were off. She could see his eyes that were as brit as the blue skigh. He had brown fur and his tail was all black. Lisa had the green leash picked out and couldn't wait to bye it. This was the best birthday ever!

1. _____ 4. _____ 7. _____

2. _____ 5. _____

3. _____ 6. _____

Writing Activity

Write about something you can't wait to happen. Use at least three spelling words in your description.

_____

_____

_____

Name_____

Look at the words in each set below. One word in each set is spelled correctly. Look at Sample A. The letter next to the correctly spelled word in Sample A has been shaded in. Do Sample B yourself. Shade the letter of the word that is spelled correctly. When you are sure you know what to do, go on with the rest of the page.

Sample A:

Ⓐ cind
Ⓑ kind
Ⓒ kynd
Ⓓ cynd

Sample B:

Ⓔ light
Ⓕ lyte
Ⓖ liet
Ⓗ lyght

1. Ⓐ mild
   Ⓑ myld
   Ⓒ mighld
   Ⓓ mield

2. Ⓔ skie
   Ⓕ sky
   Ⓖ boale
   Ⓗ skigh

3. Ⓐ pye
   Ⓑ pigh
   Ⓒ py
   Ⓓ pie

4. Ⓔ might
   Ⓕ myte
   Ⓖ myght
   Ⓗ mayt

5. Ⓐ find
   Ⓑ fynd
   Ⓒ finde
   Ⓓ fighnd

6. Ⓔ fite
   Ⓕ fyte
   Ⓖ fight
   Ⓗ fyet

7. Ⓐ tyse
   Ⓑ tighs
   Ⓒ tiyes
   Ⓓ ties

8. Ⓔ ryte
   Ⓕ riyt
   Ⓖ right
   Ⓗ riet

9. Ⓐ frie
   Ⓑ frye
   Ⓒ fry
   Ⓓ fright

10. Ⓔ tight
    Ⓕ tite
    Ⓖ tyte
    Ⓗ tighte

11. Ⓐ chyld
    Ⓑ child
    Ⓒ childe
    Ⓓ chylde

12. Ⓔ flyte
    Ⓕ fliet
    Ⓖ flight
    Ⓗ flite

13. Ⓐ bryte
    Ⓑ bryt
    Ⓒ bryght
    Ⓓ bright

14. Ⓔ bie
    Ⓕ buy
    Ⓖ bigh
    Ⓗ buye

15. Ⓐ digh
    Ⓑ dye
    Ⓒ diegh
    Ⓓ dygh

Name_____

Read each sentence. If an underlined word is spelled wrong, fill in the circle that goes with that word. If no word is spelled wrong, fill in the circle below NONE. Read Sample A, and do Sample B.

NONE

A. She <u>sold</u> her <u>pye</u> at the <u>May</u> fair.
    A   B   C

A. Ⓐ Ⓑ Ⓒ Ⓓ

NONE

B. I <u>might</u> <u>shut</u> my eyes if the <u>skie</u> gets too bright.
      E   F   G

B. Ⓔ Ⓕ Ⓖ Ⓗ

NONE

1. There was a <u>great</u> <u>claap</u> of thunder in the <u>sky</u>.
              A   B   C

1. Ⓐ Ⓑ Ⓒ Ⓓ

NONE

2. The <u>wise</u> man <u>rose</u> early to <u>pante</u>.
          E   F   G

2. Ⓔ Ⓕ Ⓖ Ⓗ

NONE

3. I will get a good <u>grade</u> if I <u>kum</u> up with the <u>right</u> answer.
                    A   B   C

3. Ⓐ Ⓑ Ⓒ Ⓓ

NONE

4. The <u>child</u> <u>plays</u> with finger <u>paint</u>.
          E   F   G

4. Ⓔ Ⓕ Ⓖ Ⓗ

NONE

5. I <u>fail</u> to see how you could <u>myss</u> all three <u>plays</u>.
        A   B   C

5. Ⓐ Ⓑ Ⓒ Ⓓ

NONE

6. Her <u>head</u> <u>myght</u> ache if you <u>shut</u> the door too loudly.
          E   F   G

6. Ⓔ Ⓕ Ⓖ Ⓗ

NONE

7. The <u>lofe</u> of bread and the <u>pie</u> were <u>slow</u> to bake.
          A   B   C

7. Ⓐ Ⓑ Ⓒ Ⓓ

NONE

8. In <u>May</u> we will <u>come</u> to Boston to visit my <u>grait</u> aunt.
        E   F   G

8. Ⓔ Ⓕ Ⓖ Ⓗ

NONE

9. The balloon <u>rose</u> high into the <u>golde</u> <u>sky</u>.
                  A   B   C

9. Ⓐ Ⓑ Ⓒ Ⓓ

NONE

10. The <u>chyld</u> will <u>clap</u> and <u>smile</u> when her mother returns.
            E   F   G

10. Ⓔ Ⓕ Ⓖ Ⓗ

NONE

11. She was <u>slow</u> to admit she <u>might</u> <u>fale</u> the test.
              A   B   C

11. Ⓐ Ⓑ Ⓒ Ⓓ

12. My uncle <u>grows</u> a <u>crop</u> of <u>gold</u> wheat on his farm.
             E   F   G

NONE
12. Ⓔ Ⓕ Ⓖ Ⓗ

13. I <u>miss</u> seeing the <u>loaf</u> of bread as it <u>gros</u> in the oven.
      A   B   C

NONE
13. Ⓐ Ⓑ Ⓒ Ⓓ

14. It is not <u>rihte</u> to <u>smile</u> when others <u>fail</u>.
           E   F   G

NONE
14. Ⓔ Ⓕ Ⓖ Ⓗ

15. She <u>sold</u> the red <u>roze</u> to the <u>wise</u> woman.
      A   B   C

NONE
15. Ⓐ Ⓑ Ⓒ Ⓓ

16. The <u>head</u> of the school <u>schut</u> the door to talk to the
       E   F
<u>grade</u> 4 teacher.
 G

NONE
16. Ⓔ Ⓕ Ⓖ Ⓗ

17. The corn <u>crop</u> <u>grows</u> at a <u>slo</u> rate.
          A   B   C

NONE
17. Ⓐ Ⓑ Ⓒ Ⓓ

18. We used <u>gold</u> <u>paint</u> on the costumes for the <u>plaise</u>.
           E F   G

NONE
18. Ⓔ Ⓕ Ⓖ Ⓗ

19. She will <u>miss</u> the apple <u>piy</u> if she does not <u>come</u> soon.
         A   B   C

NONE
19. Ⓐ Ⓑ Ⓒ Ⓓ

20. The <u>child</u> will finish third <u>graid</u> in <u>May</u>.
       E   F   G

NONE
20. Ⓔ Ⓕ Ⓖ Ⓗ

21. It is <u>wyze</u> to <u>shut</u> the windows if the <u>sky</u> turns gray.
      A   B   C

NONE
21. Ⓐ Ⓑ Ⓒ Ⓓ

22. The <u>right</u> <u>crop</u> can always be <u>solde</u>.
      E F   G

NONE
22. Ⓔ Ⓕ Ⓖ Ⓗ

23. The <u>loaf</u> was <u>sold</u> to the man at the <u>hade</u> table.
      A   B   C

NONE
23. Ⓐ Ⓑ Ⓒ Ⓓ

24. We <u>clap</u> when the <u>great</u> actors <u>smile</u>.
      E   F   G

NONE
24. Ⓔ Ⓕ Ⓖ Ⓗ

25. I <u>might</u> bake a different <u>pie</u> this <u>Mai</u>.
     A   B   C

NONE
25. Ⓐ Ⓑ Ⓒ Ⓓ

Name _____

Fold back the paper along the dotted line. Use the blanks to write each word as it is read aloud. When you finish the test, unfold the paper. Use the list at the right to correct any spelling mistakes.

1. _____
2. _____
3. _____
4. _____
5. _____
6. _____
7. _____
8. _____
9. _____
10. _____
11. _____
12. _____
13. _____
14. _____
15. _____

Review Words

16. _____
17. _____
18. _____

Challenge Words

19. _____
20. _____

1. heel
2. seal
3. weak
4. week
5. bean
6. creek
7. speaks
8. team
9. free
10. green
11. clean
12. cream
13. street
14. freeze
15. field
16. right
17. pie
18. child
19. sixteen
20. peanut

At Home: Help your child practice the words he or she missed to prepare for the Posttest.

The Strongest One • Book I/Unit 2   33

Name_____

Using the Word Study Steps

1. LOOK at the word.
2. SAY the word aloud.
3. STUDY the letters in the word.

4. WRITE the word.
5. CHECK the word.
   Did you spell the word right?
   If not, go back to step 1.

Crossword Puzzle

Solve the crossword puzzle with spelling words that complete the sentences.

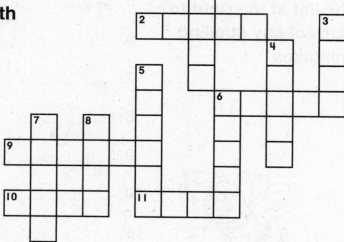

ACROSS

2. After the show, we _____ the dressing rooms.
6. There is a scene in the play that happens near a watery _____.
9. If you leave the water outside, it will _____.
10. James played the part of a baby _____ in the zoo scene.
11. I clap when my _____ scores a goal.

DOWN

1. The ants carried a small black _____ away from the picnic.
3. I was too _____ to carry the heavy sets.
4. We passed a _____ that was lit by a beautiful sunset.
5. Sometimes we play our guitars on the _____ to make extra money.
6. Aunt Sarah likes to put _____ in her tea.
7. I painted the leaves on the forest set bright _____.
8. Some gum got stuck under the _____ of my shoe.

© Macmillan/McGraw-Hill

At Home: Review the Word Study Steps to help your child spell new words.

Name_____

heel	week	speaks	green	street
seal	bean	team	clean	freeze
weak	creek	free	cream	field

Vowel Power

Write the spelling words that contain the matching spelling of the long e sound.

long *e* spelled *ea*

1. _____   5. _____

2. _____   6. _____

3. _____   7. _____

4. _____

long *e* spelled *ee*

8. _____   12. _____

9. _____   13. _____

10. _____   14. _____

11. _____

Rhyme Time

Write the spelling word that rhymes with each word below.

15. treat _____

16. bean _____

Name_____

heel	week	speaks	green	street
seal	bean	team	clean	freeze
weak	creek	free	cream	field

It Takes Three

Write a spelling word that goes with the other two words.

1. blue, yellow, _____

2. pea, pod, _____

3. group, club, _____

4. road, highway, _____

What Does It Mean?

Write a spelling word that matches each clue below.

5. Says   _____

6. Thick milk   _____

7. A grassy area   _____

8. Not strong   _____

9. Make ice from water   _____

10. Scrub or wash   _____

11. A part of the foot   _____

12. Close tightly   _____

13. Small stream   _____

14. At no cost   _____

15. Seven days   _____

Name_____

Proofreading

There are six spelling mistakes in these paragraphs. Circle the misspelled words. Write the words correctly on the lines below.

This year I joined the dance teem at school. It's been so much fun. We've been practicing for a long time for our big show. I play the part of a butterfly. I get to do one dance all by myself.

I can't believe that it is just one weke away. The teachers helped us build our stage outside on the soccer feeld. My mom made my costume. It's dark grean with pretty, colorful wings. Mom also fixed the loose heal of my shoe.

Mom and dad bought their tickets last week. My sister gets hers for frie because she goes to school here, too. Some of the other dancers are nervous, but not me. I can't wait for the show to begin.

1. _____   4. _____

2. _____   5. _____

3. _____   6. _____

Writing Activity

Write a description of a favorite after-school activity. Use at least four spelling words in your description.

_____

_____

_____

_____

_____

Name_____

Look at the words in each set below. One word in each set is spelled correctly. Look at Sample A. The letter next to the correctly spelled word in Sample A has been shaded in. Do Sample B yourself. Shade the letter of the word that is spelled correctly. When you are sure you know what to do, go on with the rest of the page.

Sample A:

- Ⓐ grene
- Ⓑ grean
- Ⓒ grien
- Ⓓ green

Sample B:

- Ⓔ sneze
- Ⓕ sneeze
- Ⓖ sneez
- Ⓗ sneaze

1. Ⓐ heale
   Ⓑ hele
   Ⓒ heel
   Ⓓ hiel

2. Ⓔ siel
   Ⓕ seel
   Ⓖ sele
   Ⓗ seal

3. Ⓐ weik
   Ⓑ weeke
   Ⓒ weake
   Ⓓ weak

4. Ⓔ wiek
   Ⓕ week
   Ⓖ weik
   Ⓗ weake

5. Ⓐ bean
   Ⓑ bene
   Ⓒ beane
   Ⓓ beene

6. Ⓔ criek
   Ⓕ creake
   Ⓖ creek
   Ⓗ creke

7. Ⓐ speeks
   Ⓑ speaks
   Ⓒ speiks
   Ⓓ spekes

8. Ⓔ tiem
   Ⓕ teim
   Ⓖ team
   Ⓗ teme

9. Ⓐ free
   Ⓑ frea
   Ⓒ frey
   Ⓓ frie

10. Ⓔ grene
    Ⓕ green
    Ⓖ grean
    Ⓗ greene

11. Ⓐ clene
    Ⓑ clean
    Ⓒ cleen
    Ⓓ cleene

12. Ⓔ creem
    Ⓕ creim
    Ⓖ criem
    Ⓗ cream

13. Ⓐ streat
    Ⓑ street
    Ⓒ streit
    Ⓓ striet

14. Ⓔ freeze
    Ⓕ freze
    Ⓖ freaz
    Ⓗ freaze

15. Ⓐ feeld
    Ⓑ field
    Ⓒ feald
    Ⓓ feild

© Macmillan/McGraw-Hill

Name _____

Fold back the paper along the dotted line. Use the blanks to write each word as it is read aloud. When you finish the test, unfold the paper. Use the list at the right to correct any spelling mistakes.

1. _____
2. _____
3. _____
4. _____
5. _____
6. _____
7. _____
8. _____
9. _____
10. _____
11. _____
12. _____
13. _____
14. _____
15. _____

Review Words 16. _____

17. _____

18. _____

Challenge Words 19. _____

20. _____

1. chick
2. much
3. pitch
4. teacher
5. chum
6. lunch
7. ditch
8. cheek
9. hatch
10. cheese
11. bench
12. chunk
13. stretch
14. watching
15. crunching
16. weak
17. green
18. seal
19. catcher
20. sandwich

At Home: Help your child practice the words he or she missed to prepare for the Posttest.

Name _____

Using the Word Study Steps

1. LOOK at the word.

2. SAY the word aloud.

3. STUDY the letters in the word.

4. WRITE the word.

5. CHECK the word.
   Did you spell the word right?
   If not, go back to step 1.

Find and Circle

Where are the spelling words?

C	H	E	E	K	W	L	K	S	Y
C	H	U	N	K	A	M	G	T	T
H	A	X	Y	R	T	V	B	R	E
E	C	H	U	M	C	P	E	E	A
E	D	I	T	C	H	I	N	T	C
S	Q	W	H	E	I	T	C	C	H
E	F	J	L	U	N	C	H	H	E
V	M	U	C	H	G	H	E	F	R
C	R	U	N	C	H	I	N	G	S
H	A	T	C	H	C	H	I	C	K

At Home: Review the Word Study Steps to help your child spell new words.

Name_____

chick	teacher	ditch	cheese	stretch
much	chum	cheek	bench	watching
pitch	lunch	hatch	chunk	crunching

Write the spelling words with these spelling patterns.

ch

1. _____
2. _____
3. _____
4. _____
5. _____

6. _____
7. _____
8. _____
9. _____
10. _____

tch

11. _____
12. _____
13. _____

14. _____
15. _____

Sounds Alike

Write the spelling words that rhyme with the words below. Then circle the letters that spell the /ch/ sound in each word.

16. leak _____
17. clench _____
18. lick _____

19. munch _____
20. such _____

Name _____

chick	teacher	ditch	cheese	stretch
much	chum	cheek	bench	watching
pitch	lunch	hatch	chunk	crunching

What's the Word?

Complete each sentence with a word from the spelling list.

1. The snow made a _____ sound as the wolf walked across it.

2. We eat _____ after science class.

3. When will the birds _____ from their eggs?

4. I sat down on a park _____ to write my letter.

5. The baby _____ was soft and fuzzy.

6. Our _____ told us a story about a wolf.

7. He would like a grilled _____ sandwich.

8. Are you _____ the baseball game on Friday night?

9. _____ the ball over the plate.

10. You should take a walk to _____ your legs.

Define It!

Write the spelling words that have the same meaning as the words or phrases below.

11. friend _____

12. a lot _____

13. large piece _____

14. part of a face _____

15. hole _____

Name_____

Proofreading

**There are six spelling mistakes in the story. Circle the
misspelled words. Write the words correctly on the lines below.**

One day a wolf went out in the woods for a walk. He was hungry
so he went looking for some lounch. He saw a dich along the path. He
decided to hide there to wait for a nice fat mouse to come along.

The wolf sat in the hole for a long time. He was tired of waching,
but he was still very hungry. Suddenly, he heard footsteps cruncing the
leaves on the path. He peeked out of the ditch and saw a little gray mouse
dragging a bag.

The wolf leapt out of the hole and stood in front of the little mouse.
Not very big, thought the wolf, but she will have to do. The mouse was
scared, but then she had an idea.

"How about some cheez?" she said to the wolf. She reached into her bag
and pulled out some cheese. The wolf looked at the large chunc of cheddar
that the little mouse offered and then at the tiny size of the mouse.

"Thank you," said the wolf. "That would be great."

1. _____   3. _____   5. _____

2. _____   4. _____   6. _____

Writing Activity

**Write a story about a two animals meeting in the woods. Use
four spelling words in your writing.**

_____

_____

_____

_____

_____

Name_____

Look at the words in each set below. One word in each set is spelled correctly. Look at Sample A. The letter next to the correctly spelled word in Sample A has been shaded in. Do Sample B yourself. Shade the letter of the word that is spelled correctly. When you are sure you know what to do, go on with the rest of the page.

Sample A:

- Ⓐ sheck
- Ⓑ ceck
- Ⓒ check
- Ⓓ schek

Sample B:

- Ⓔ watch
- Ⓕ wach
- Ⓖ watsch
- Ⓗ whach

1. Ⓐ chick
   Ⓑ shick
   Ⓒ chik
   Ⓓ schick

2. Ⓔ mutch
   Ⓕ mutsh
   Ⓖ musch
   Ⓗ much

3. Ⓐ pish
   Ⓑ pitsh
   Ⓒ pitch
   Ⓓ pich

4. Ⓔ techer
   Ⓕ teacher
   Ⓖ teasher
   Ⓗ teatcher

5. Ⓐ chumm
   Ⓑ schum
   Ⓒ chume
   Ⓓ chum

6. Ⓔ lonch
   Ⓕ lunsh
   Ⓖ lunch
   Ⓗ lunsch

7. Ⓐ ditsh
   Ⓑ dich
   Ⓒ detch
   Ⓓ ditch

8. Ⓔ sheak
   Ⓕ cheek
   Ⓖ cheak
   Ⓗ cheec

9. Ⓐ hach
   Ⓑ hatch
   Ⓒ hasch
   Ⓓ hatsch

10. Ⓔ cheese
    Ⓕ cheeze
    Ⓖ sheese
    Ⓗ chese

11. Ⓐ bench
    Ⓑ bentch
    Ⓒ bensh
    Ⓓ bentsh

12. Ⓔ shunck
    Ⓕ shunk
    Ⓖ chunck
    Ⓗ chunk

13. Ⓐ strech
    Ⓑ stretch
    Ⓒ stretsch
    Ⓓ shtrech

14. Ⓔ waching
    Ⓕ wasching
    Ⓖ watching
    Ⓗ wathing

15. Ⓐ crunching
    Ⓑ cruntching
    Ⓒ cruching
    Ⓓ crunshing

Name _____

Fold back the paper along the dotted line. Use the blanks to write each word as it is read aloud. When you finish the test, unfold the paper. Use the list at the right to correct any spelling mistakes.

1. _____
2. _____
3. _____
4. _____
5. _____
6. _____
7. _____
8. _____
9. _____
10. _____
11. _____
12. _____
13. _____
14. _____
15. _____

Review Words 16. _____

17. _____

18. _____

Challenge Words 19. _____

20. _____

1. thick
2. this
3. truth
4. whales
5. shock
6. fish
7. what
8. sixth
9. them
10. washing
11. wheel
12. pathway
13. month
14. dishpan
15. weather
16. lunch
17. chick
18. pitch
19. shadow
20. thicken

At Home: Help your child practice the words he or she missed to prepare for the Posttest.

What's in Store for the Future 45
Book I/Unit 2

Name_____

Using the Word Study Steps

1. LOOK at the word.

2. SAY the word aloud.

3. STUDY the letters in the word.

4. WRITE the word.

5. CHECK the word.
   Did you spell the word right?
   If not, go back to step 1.

Find Rhyming Words

Circle the word in each row that rhymes with the spelling word on the left.

1. **snails**	whales	smells	walls
2. **seal**	sell	wheel	swell
3. **feather**	farther	weather	whisper
4. **stick**	thick	stink	sink
5. **tooth**	soot	toot	truth
6. **miss**	mister	this	think
7. **wish**	fish	fit	finish
8. **shut**	shot	shop	what
9. **stem**	stern	storm	them
10. **rock**	shake	shock	short

© Macmillan/McGraw-Hill

 At Home: Review the Word Study Steps to help your child spell new words.

Name_____

thick	whales	what	washing	month
this	shock	sixth	wheel	dishpan
truth	fish	them	pathway	weather

Write the spelling words with these spelling patterns.

th

1. _____   4. _____   7. _____

2. _____   5. _____   8. _____

3. _____   6. _____

wh   *sh*

9. _____   12. _____

10. _____   13. _____

11. _____   14. _____

15. _____

Syllable Stuff

Write the spelling words that have one syllable:

16. _____   20. _____   24. _____

17. _____   21. _____   25. _____

18. _____   22. _____   26. _____

19. _____   23. _____

Write the spelling words that have two syllables:

27. _____   29. _____

28. _____   30. _____

Name_____

thick	whales	what	washing	month
this	shock	sixth	wheel	dishpan
truth	fish	them	pathway	weather

What's the Word?

Complete each sentence with a spelling word.

1. In college, I will study _____ events such as hurricanes.

2. She was _____ her hands before cooking class.

3. Next _____ my family will go visit my aunt.

4. I want to study _____, fish, and other ocean life when I grow up.

5. Last year, Mary came in _____ place in the national spelling bee.

6. The _____ is that no one knows what will happen many years from now.

7. _____ book is very interesting.

8. _____ do you want to be when you grow up?

9. People who lived in the past are interesting. I like to learn about _____.

10. Eating _____ has always been part of the diet in the United States.

Define It!

Write the spelling words that have the same meaning as the words or phrases below.

11. part of a car   _____

12. tub for washing plates   _____

13. upset   _____

14. trail   _____

Name _____

Proofreading

There are six spelling mistakes in the letter below. Circle the misspelled words. Write the words correctly on the lines below.

Dear Joe,

    I have to tell you about my trip to the coast last munth. We went out on a boat to look for wales! It was quite an adventure. It made me very happy because I want to study ocean life when I grow up.

    The wether was nice when we set out, sunny and warm. We saw a lot of fisch, which was fun. I took many notes. After we ate lunch, the sky got very dark and it started to rain. The boat was big but it still rocked a lot in the waves. To tell you the trouth, I was a little bit scared.

    Then, we saw thiem: ten whales swimming in the distance. It was great.

    When I get older, I want to learn more about these animals. Our guide told me that I could study whales in college. When I get my photos back, I will send you one.

                 Your friend,

                 Elizabeth

1. _____   3. _____   5. _____

2. _____   4. _____   6. _____

Writing Activity

Write a paragraph about what kind of job you would like to have in the future. Use at least four spelling words.

_____

_____

_____

_____

Name_____

Look at the words in each set below. One word in each set
is spelled correctly. Look at Sample A. The letter next to the
correctly spelled word in Sample A has been shaded in. Do
Sample B yourself. Shade the letter of the word that is spelled
correctly. When you are sure you know what to do, go on with
the rest of the page.

Sample A:

Ⓐ each
Ⓑ eatch
Ⓒ eash
Ⓓ eatsch

Sample B:

Ⓔ wich
Ⓕ weth
Ⓕ wit
Ⓗ with

1. Ⓐ thisk
   Ⓑ tick
   Ⓒ thich
   Ⓓ thick

2. Ⓔ thiss
   Ⓕ this
   Ⓖ thes
   Ⓗ thess

3. Ⓐ truuth
   Ⓑ trut
   Ⓒ truth
   Ⓓ ruth

4. Ⓔ wales
   Ⓕ whayles
   Ⓖ whales
   Ⓗ whaels

5. Ⓐ shawk
   Ⓑ shock
   Ⓒ schawk
   Ⓓ chack

6. Ⓔ feesh
   Ⓕ fish
   Ⓖ fich
   Ⓗ fesh

7. Ⓐ what
   Ⓑ wat
   Ⓒ watt
   Ⓓ whaut

8. Ⓔ sith
   Ⓕ sickth
   Ⓖ siskth
   Ⓗ sixth

9. Ⓐ them
   Ⓑ tem
   Ⓒ thum
   Ⓓ dem

10. Ⓔ waching
    Ⓕ washing
    Ⓖ whashing
    Ⓗ wasching

11. Ⓐ wheel
    Ⓑ weel
    Ⓒ wheal
    Ⓓ while

12. Ⓔ patway
    Ⓕ pathwhay
    Ⓖ phatway
    Ⓗ pathway

13. Ⓐ mont
    Ⓑ munth
    Ⓒ muntth
    Ⓓ month

14. Ⓔ dishpan
    Ⓕ dichpan
    Ⓖ ditchpan
    Ⓗ deshpan

15. Ⓐ wether
    Ⓑ wheather
    Ⓒ weather
    Ⓓ weater

Name _____

Fold back the paper along the dotted line. Use the blanks to write each word as it is read aloud. When you finish the test, unfold the paper. Use the list at the right to correct any spelling mistakes.

1. _____
2. _____
3. _____
4. _____
5. _____
6. _____
7. _____
8. _____
9. _____
10. _____
11. _____
12. _____
13. _____
14. _____
15. _____

Review Words

16. _____
17. _____
18. _____

Challenge Words

19. _____
20. _____

1. thread
2. scrubs
3. spree
4. screams
5. stream
6. scratch
7. spread
8. throne
9. three
10. screens
11. spray
12. throw
13. strong
14. scraped
15. strength
16. thick
17. washing
18. whales
19. streamer
20. scribble

At Home: Help your child practice the words he or she missed to prepare for the Posttest.

The Planets in Our Solar System
Book I/Unit 2
51

Name _____

Using the Word Study Steps

1. LOOK at the word.

2. SAY the word aloud.

3. STUDY the letters in the word.

4. WRITE the word.

5. CHECK the word.
   Did you spell the word right?
   If not, go back to step 1.

Find and Circle

Where are the spelling words?

P	C	S	C	R	E	E	N	S	W	X
S	T	R	E	N	G	T	H	P	Z	K
C	S	B	M	F	J	H	U	R	V	T
R	C	W	S	T	H	R	E	A	D	H
U	R	S	P	L	Z	O	N	Y	Y	R
B	A	C	R	R	M	W	G	K	P	O
S	P	R	E	E	O	O	Q	W	N	N
S	E	E	A	S	T	R	O	N	G	E
H	D	A	D	D	T	H	R	E	E	Z
A	Q	M	H	S	T	R	E	A	M	U
F	V	S	C	R	A	T	C	H	S	C

At Home: Review the Word Study Steps to help your child
spell new words.

Name_____

thread	screams	spread	screens	strong
scrubs	stream	throne	spray	scraped
spree	scratch	three	throw	strength

Write the spelling words for each of these clusters below.

thr

1. _____
2. _____
3. _____
4. _____

scr

5. _____
6. _____
7. _____
8. _____
9. _____

spr

10. _____
11. _____
12. _____

str

13. _____
14. _____
15. _____

What's in a Word?

Write the spelling words in which you can find the smaller word.

16. ray   _____
17. rub   _____
18. row   _____

Name_____

thread	screams	spread	screens	strong
scrubs	stream	throne	spray	scraped
spree	scratch	three	throw	strength

It Takes Three

Write a spelling word that goes with the other two words.

1. sewing, needle, _____

2. yells, hollers, _____

3. river, brook, _____

4. mist, water, _____

5. cleans, washes, _____

What Does It Mean?

Write a spelling word that matches each clue below.

6. Where a king sits _____

7. Having great power _____

8. What they show movies on _____

9. What you do with a ball _____

10. To cut with a fingernail _____

Make a Sentence

Use each word in a sentence.

11. strong _____

12. three _____

13. scraped _____

14. spree _____

Name_____

Proofreading

There are six spelling mistakes in this TV broadcast. Circle the misspelled words. Write the words correctly on the lines below.

Hello, everyone! This is Wendy Mills reporting live from the space shuttle lift-off. And what a sight it is. The shuttle is ready to go. The crowd is spreed out across a huge field. You can hear sckreems of excitement all around. NASA has set up three big schrenes that show what is going on inside the shuttle. The captain sits in a chair that is as big as a threwn! Outside, workers are lifting lots of heavy equipment. Wow, they sure must be stroon to carry all that weight. Any minute now they will thro the switch and the shuttle will take off. Stay tuned.

1. _____   3. _____   5. _____

2. _____   4. _____   6. _____

Writing Activity

Write a news report of an event you have seen. Use at least four spelling words in your description.

_____

_____

_____

_____

_____

_____

_____

Name_____

Look at the words in each set below. One word in each set is spelled correctly. Look at Sample A. The letter next to the correctly spelled word in Sample A has been shaded in. Do Sample B yourself. Shade the letter of the word that is spelled correctly. When you are sure you know what to do, go on with the rest of the page.

Sample A:

Ⓐ trea
Ⓑ three
Ⓒ thee
Ⓓ threa

Sample B:

Ⓔ skrape
Ⓕ scraep
Ⓖ scrape
Ⓗ scrapp

1. Ⓐ thred
   Ⓑ thead
   Ⓒ thread
   Ⓓ thraed

2. Ⓔ scrubs
   Ⓕ skrubs
   Ⓖ scrubes
   Ⓗ scubs

3. Ⓐ sprea
   Ⓑ spre
   Ⓒ shree
   Ⓓ spree

4. Ⓔ skreams
   Ⓕ screams
   Ⓖ sreems
   Ⓗ screims

5. Ⓐ steem
   Ⓑ streem
   Ⓒ sream
   Ⓓ stream

6. Ⓔ scatch
   Ⓕ shratch
   Ⓖ scratch
   Ⓗ scratsch

7. Ⓐ shpread
   Ⓑ spread
   Ⓒ shread
   Ⓓ spred

8. Ⓔ throne
   Ⓕ thrown
   Ⓖ trown
   Ⓗ trauwn

9. Ⓐ threa
   Ⓑ tree
   Ⓒ thre
   Ⓓ three

10. Ⓔ screens
    Ⓕ screans
    Ⓖ skreens
    Ⓗ skreams

11. Ⓐ spray
    Ⓑ scray
    Ⓒ sprae
    Ⓓ spraiy

12. Ⓔ throw
    Ⓕ thrauw
    Ⓖ trow
    Ⓗ throwe

13. Ⓐ shtrong
    Ⓑ strong
    Ⓒ stron
    Ⓓ stronge

14. Ⓔ scraped
    Ⓕ scrapped
    Ⓖ scapped
    Ⓗ skraped

15. Ⓐ stength
    Ⓑ strangth
    Ⓒ strength
    Ⓓ strenth

Name_____

Fold back the paper along the dotted line. Use the blanks to write each word as it is read aloud. When you finish the test, unfold the paper. Use the list at the right to correct any spelling mistakes.

1. _____
2. _____
3. _____
4. _____
5. _____
6. _____
7. _____
8. _____
9. _____
10. _____
11. _____
12. _____
13. _____
14. _____
15. _____

Review Words
16. _____
17. _____
18. _____

Challenge Words
19. _____
20. _____

1. wrap
2. knit
3. gnat
4. wrists
5. knots
6. wrote
7. knight
8. sign
9. knock
10. wreck
11. know
12. wring
13. gnaws
14. write
15. wrong
16. throw
17. spray
18. scratch
19. wristwatch
20. knapsack

At Home: Help your child practice the words he or she missed to prepare for the Posttest.

A True Story • Book I/Unit 2   57

Name_____

Using the Word Study Steps

1. LOOK at the word.

2. SAY the word aloud.

3. STUDY the letters in the word.

4. WRITE the word.

5. CHECK the word.
Did you spell the word right?
If not, go back to step 1.

Find Rhyming Words

Circle the word in each row that rhymes with the word in dark type.

1. **trap**	rip	tarp	wrap
2. **float**	foam	flat	wrote
3. **song**	wrong	sing	roam
4. **sight**	style	sit	knight
5. **rock**	known	knock	roll
6. **mine**	sign	mend	mint
7. **flaws**	gnaws	flame	naps
8. **sing**	wring	sang	write
9. **bit**	bite	knit	bait
10. **deck**	dock	deal	wreck
11. **sat**	gnat	gale	still
12. **glow**	glee	know	blew
13. **mists**	mast	gusts	wrists
14. **blots**	blast	knots	cost
15. **sight**	write	seen	shell

© Macmillan/McGraw-Hill

At Home: Review the Word Study Steps to help your child spell new words.

Name_____

wrap	wrists	knight	wreck	gnaws
knit	knots	sign	know	write
gnat	wrote	knock	wring	wrong

Pattern Power!

Write the spelling words that have each silent letter.

w

1. _____   5. _____

2. _____   6. _____

3. _____   7. _____

4. _____

k   **g**

8. _____   13. _____

9. _____   14. _____

10. _____   15. _____

11. _____

12. _____

Word Hunt

Write the spelling words in which each of the words below can be found.

16. night _____   19. now _____

17. not _____   20. rap _____

18. ring _____

Name_____

wrap	wrists	knight	wreck	gnaws
knit	knots	sign	know	write
gnat	wrote	knock	wring	wrong

Part of the Group

Read the heading for each group of words. Then add the spelling word that belongs in each group.

Things you do with thread

1. sew, stitch, _____

Parts of arms

2. hands, elbows, _____

Bugs

3. fly, spider, _____

Things you tie

4. bows, shoelaces, _____

In the Dictionary

Many dictionary entries have sample sentences that show how the word can be used. Complete each sample sentence with a spelling word.

5. The _____ at the library said "Story Hour."

6. Someday I want to _____ the story of my life.

7. Marcy did not _____ her family history.

8. This is a fairy tale about a monster that _____ through stone.

Opposites!

Write the spelling word that has the opposite meaning.

9. right _____

10. fix _____

Name_____

Proofreading

There are five spelling mistakes in this letter. Circle the misspelled words. Write the words correctly on the lines below.

Dear Diary,

An author came to school today to tell us about the book she wroat. It took her two years! It is about the many adventures of a nite. First, he is on a ship. Then, he is captured by an evil band of robbers. In the end, he becomes a king.

It is exciting to noe someone who is an author. I think I want to rit a book when I grow up. It would be about something I love. Maybe it will be about a lady who likes to nit pretty scarves and hats, just like my grandmother. Or maybe it will be about about my pet dog. Who knows!

I will write more tomorrow,

Alice

1. _____   3. _____   5. _____

2. _____   4. _____

Writing Activity

Write about a book you would like to write. Use at least four spelling words in your description.

_____

_____

_____

_____

_____

_____

Name_____

Look at the words in each set below. One word in each set is spelled correctly. Look at Sample A. The letter next to the correctly spelled word in Sample A has been shaded in. Do Sample B yourself. Shade the letter of the word that is spelled correctly. When you are sure you know what to do, go on with the rest of the page.

Sample A:

Ⓐ kneal
Ⓑ neal
Ⓒ kneel
Ⓓ neel

Sample B:

Ⓔ fite
Ⓕ fight
Ⓖ fighte
Ⓗ faete

1. Ⓐ rap
   Ⓑ rapp
   Ⓒ wrapp
   Ⓓ wrap

2. Ⓔ gnit
   Ⓕ knigt
   Ⓖ knit
   Ⓗ gnite

3. Ⓐ knat
   Ⓑ gnat
   Ⓒ nat
   Ⓓ gnate

4. Ⓔ rists
   Ⓕ wrighsts
   Ⓖ rhist
   Ⓗ wrists

5. Ⓐ gnots
   Ⓑ knots
   Ⓒ knotes
   Ⓓ gnaughts

6. Ⓔ roat
   Ⓕ wroat
   Ⓖ roate
   Ⓗ wrote

7. Ⓐ knight
   Ⓑ gnite
   Ⓒ gnight
   Ⓓ knite

8. Ⓔ signe
   Ⓕ sihn
   Ⓖ sign
   Ⓗ sighn

9. Ⓐ knock
   Ⓑ gnock
   Ⓒ knawk
   Ⓓ gnack

10. Ⓔ wreck
    Ⓕ wrek
    Ⓖ reack
    Ⓗ reck

11. Ⓐ nowe
    Ⓑ know
    Ⓒ gnow
    Ⓓ knoa

12. Ⓔ wring
    Ⓕ wrign
    Ⓖ wreng
    Ⓗ wrang

13. Ⓐ knaws
    Ⓑ naws
    Ⓒ gnaus
    Ⓓ gnaws

14. Ⓔ ryte
    Ⓕ righte
    Ⓖ write
    Ⓗ wryte

15. Ⓐ rong
    Ⓑ wronge
    Ⓒ wrong
    Ⓓ ronge

Name _____

Read each sentence. If an underlined word is spelled wrong, fill in the circle that goes with that word. If no word is spelled wrong, fill in the circle below NONE. Read Sample A, and do Sample B.

A. I want to <u>see</u> the <u>whales</u> very <u>mutch</u>.
          A   B   C

                                                 NONE
A. Ⓐ Ⓑ ⓒ Ⓓ

B. The <u>truth</u> is that <u>nite</u> should be on the <u>throne</u>.
        E   F   G

                                                 NONE
B. Ⓔ Ⓕ Ⓖ Ⓗ

1. I <u>thru</u> the <u>pitch</u> as fast as I could so my <u>team</u> would win.
     A   B   C

                                                 NONE
1. Ⓐ Ⓑ ⓒ Ⓓ

2. We were <u>watching</u> the <u>wethur</u> report, when there
                   E   F

   was a <u>knock</u> at the door.
              G

                                               NONE
2. Ⓔ Ⓕ Ⓖ Ⓗ

3. We sat on a <u>clean</u> <u>beanche</u> to eat our <u>lunch</u>.
               A   B   C

                                                 NONE
3. Ⓐ Ⓑ ⓒ Ⓓ

4. The <u>green</u> <u>fish</u> swam in the <u>streeme</u>.
          E   F   G

                                                 NONE
4. Ⓔ Ⓕ Ⓖ Ⓗ

5. Her <u>teacher</u> was <u>watching</u> her <u>sine</u> her name.
         A   B   C

                                                 NONE
5. Ⓐ Ⓑ ⓒ Ⓓ

6. The boy sat on the <u>bench</u> <u>wachin</u> his <u>team</u> play.
                      E   F   G

                                                 NONE
6. Ⓔ Ⓕ Ⓖ Ⓗ

7. I <u>know</u> that <u>this</u> gum must be <u>scraped</u> off the desk.
     A   B   C

                                                 NONE
7. Ⓐ Ⓑ ⓒ Ⓓ

8. The cook <u>scrubs</u> the pots after <u>lunch</u> once a <u>wekk</u>.
             E   F   G

                                                 NONE
8. Ⓔ Ⓕ Ⓖ Ⓗ

9. Her <u>teacher</u> put her on the <u>green</u> <u>teeme</u>.
     A   B   C

                                                 NONE
9. Ⓐ Ⓑ ⓒ Ⓓ

NONE

10. I <u>threw</u> a <u>strong</u> <u>peich</u>, but I did not mean to break   10. Ⓔ Ⓕ Ⓖ Ⓗ
      E   F   G

the window.

NONE

11. Last <u>week</u> my sister <u>skrapped</u> her knee when she   11. Ⓐ Ⓑ Ⓒ Ⓓ
       A   B

fell on the <u>street</u>.
         C

NONE

12. Once a <u>munthe</u> they <u>clean</u> the <u>street</u>.   12. Ⓔ Ⓕ Ⓖ Ⓗ
       E   F   G

NONE

13. There were <u>knots</u> in the <u>thick</u> yarn I used to <u>knit</u> the   13. Ⓐ Ⓑ Ⓒ Ⓓ
        A   B   C

sweater.

NONE

14. I <u>noew</u> how to <u>sign</u> my name on <u>this</u> paper.   14. Ⓔ Ⓕ Ⓖ Ⓗ
      E   F   G

NONE

15. Joe asked his coach <u>what</u> was <u>rong</u> with the <u>pitch</u>.   15. Ⓐ Ⓑ Ⓒ Ⓓ
         A   B   C

NONE

16. The nurse <u>skrubbs</u> his <u>scraped</u> leg, while he stares   16. Ⓔ Ⓕ Ⓖ Ⓗ
       E   F

at the <u>stream</u> outside his window.
      G

NONE

17. The <u>weather</u> <u>theiss</u> <u>month</u> has been very mild.   17. Ⓐ Ⓑ Ⓒ Ⓓ
      A   B   C

NONE

18. <u>Sign</u> in, <u>write</u> your address, then <u>nokk</u> on that door.   18. Ⓔ Ⓕ Ⓖ Ⓗ
    E   F   G

NONE

19. You <u>know</u> how she <u>scrubs</u> the floor until it is <u>klene</u>.   19. Ⓐ Ⓑ Ⓒ Ⓓ
      A   B   C

NONE

20. <u>Watt</u> kind of <u>fish</u> would you like for <u>lunch</u>?   20. Ⓔ Ⓕ Ⓖ Ⓗ
    E   F   G

© Macmillan/McGraw-Hill

Name _____

Fold back the paper along the dotted line. Use the blanks to write each word as it is read aloud. When you finish the test, unfold the paper. Use the list at the right to correct any spelling mistakes.

1. _____
2. _____
3. _____
4. _____
5. _____
6. _____
7. _____
8. _____
9. _____
10. _____
11. _____
12. _____
13. _____
14. _____
15. _____

Review Words 16. _____
17. _____
18. _____

Challenge Words 19. _____
20. _____

1. bark
2. shorts
3. sharp
4. sore
5. hard
6. storms
7. yard
8. sport
9. sharks
10. porch
11. pour
12. story
13. chore
14. wore
15. carve
16. knots
17. sign
18. wrong
19. orchard
20. artist

At Home: Help your child practice the words he or she missed to prepare for the Posttest.

Name_____

Using the Word Study Steps

1. LOOK at the word.

2. SAY the word aloud.

3. STUDY the letters in the word.

4. WRITE the word.

5. CHECK the word.
   Did you spell the word right?
   If not, go back to step 1.

Find and Circle

Where are the spelling words?

S	H	A	R	K	S	B	S	B	N	Z
P	O	R	C	H	T	U	T	H	M	S
O	U	U	I	Z	O	P	O	U	R	H
R	Y	X	S	O	R	E	R	P	Y	O
T	Q	N	B	S	M	W	Y	C	A	R
V	C	B	V	D	S	Q	D	A	R	T
S	H	A	R	P	H	H	A	R	D	S
Z	Q	R	C	L	C	E	R	V	T	A
F	E	K	A	C	H	O	R	E	B	E
O	W	W	O	R	E	L	J	J	N	L

At Home: Review the Word Study Steps to help your child
spell new words.

Name_____

bark	sore	yard	porch	chore
shorts	hard	sport	pour	wore
sharp	storms	sharks	story	carve

Pattern Power!

This week's spelling words contain the vowel sounds /är/ and /ôr/.
Write each spelling word under the word that has the same vowel sound.

harm

1. _____   4. _____

2. _____   5. _____

3. _____   6. _____

port

7. _____   12. _____

8. _____   13. _____

9. _____   14. _____

10. _____   15. _____

11. _____

Write the spelling words with the /ôr/ sound spelled:

or   *ore*

16. _____   21. _____

17. _____   22. _____

18. _____   23. _____

19. _____

20. _____   *our*

                        24. _____

Name_____

bark	sore	yard	porch	chore
shorts	hard	sport	pour	wore
sharp	storms	sharks	story	carve

Synonym Alert!

Write the spelling words that have the same meanings as the words below.

1. pointed _____

2. tale _____

3. firm _____

4. game _____

5. shout _____

What's the Word?

Complete each sentence with a spelling word.

6. At the feast, the chef will _____ a turkey.

7. It is my _____ to set the table.

8. Last night we ate dinner outside on the _____.

9. _____ water into each glass.

10. I bought new _____ to wear.

11. _____ love to eat small fish.

12. We ate our food outside in the _____.

13. The _____ made the lights go out.

14. For the dinner party, Meg _____ her new shirt.

15. My arms were _____ after stirring the soup for 2 hours.

© Macmillan/McGraw-Hill

There are six spelling mistakes in this paragraph. Circle the misspelled words. Write the words correctly on the lines below.

Everyone in my family has a choar at dinner time. Tonight we are having a big turkey dinner on the portch. We all have a lot of work to do. Mom is in charge of the gravy. She has to open a big jar of gravy and then heat it on the stove. My job is to set the table. I put out the plates and napkins. I also porr milk into all the glasses. My big brother Mike is in charge of cutting the bread. He has to use a shorpp knife. Dad has the best job of all. It is hord, but he loves it. He has to karve the turkey!

1. _____

2. _____

3. _____

4. _____

5. _____

6. _____

Writing Activity

Write about a chore you must do each day. Use at least three spelling words in your description.

_____

_____

_____

_____

_____

_____

_____

_____

Name_____

Look at the words in each set below. One word in each set is spelled correctly. Look at Sample A. The letter next to the correctly spelled word in Sample A has been shaded in. Do Sample B yourself. Shade the letter of the word that is spelled correctly. When you are sure you know what to do, go on with the rest of the page.

Sample A:

Ⓐ erm
Ⓑ arm
Ⓒ orm
Ⓓ ahrm

Sample B:

Ⓔ corn
Ⓕ corne
Ⓖ korn
Ⓗ carn

1. Ⓐ baurk
   Ⓑ bawrk
   Ⓒ bark
   Ⓓ barke

2. Ⓔ sharts
   Ⓕ shorts
   Ⓖ sherts
   Ⓗ shurts

3. Ⓐ sharp
   Ⓑ sherp
   Ⓒ shaurp
   Ⓓ shorp

4. Ⓔ sare
   Ⓕ sor
   Ⓖ saur
   Ⓗ sore

5. Ⓐ hord
   Ⓑ harde
   Ⓒ hard
   Ⓓ haurd

6. Ⓔ sturms
   Ⓕ starms
   Ⓖ stourms
   Ⓗ storms

7. Ⓐ yard
   Ⓑ yaurd
   Ⓒ yarde
   Ⓓ yurd

8. Ⓔ sport
   Ⓕ spart
   Ⓖ spourt
   Ⓗ spowrt

9. Ⓐ shaurks
   Ⓑ shawrks
   Ⓒ sharks
   Ⓓ sherks

10. Ⓔ purch
    Ⓕ porch
    Ⓖ parche
    Ⓗ pourch

11. Ⓐ pour
    Ⓑ paur
    Ⓒ por
    Ⓓ paure

12. Ⓔ staury
    Ⓕ stary
    Ⓖ storee
    Ⓗ story

13. Ⓐ shure
    Ⓑ chure
    Ⓒ chare
    Ⓓ chore

14. Ⓔ waur
    Ⓕ wawr
    Ⓖ wore
    Ⓗ wor

15. Ⓐ carve
    Ⓑ caurve
    Ⓒ corve
    Ⓓ carv

Name _____

Fold back the paper along the dotted line. Use the blanks to write each word as it is read aloud. When you finish the test, unfold the paper. Use the list at the right to correct any spelling mistakes.

1. _____
2. _____
3. _____
4. _____
5. _____
6. _____
7. _____
8. _____
9. _____
10. _____
11. _____
12. _____
13. _____
14. _____
15. _____

Review Words 16. _____
17. _____
18. _____

Challenge Words 19. _____
20. _____

1. stairs
2. mare
3. bear
4. bare
5. share
6. wear
7. dares
8. chairs
9. glare
10. pairs
11. hare
12. their
13. pears
14. square
15. haircut
16. sport
17. sore
18. hard
19. airport
20. beware

© Macmillan/McGraw-Hill

At Home: Help your child practice the words he or she missed to prepare for the Posttest.

One Riddle, One Answer
Book I/Unit 3
71

Name_____

Using the Word Study Steps

1. LOOK at the word.

2. SAY the word aloud.

3. STUDY the letters in the word.

4. WRITE the word.

5. CHECK the word.
   Did you spell the word right?
   If not, go back to step 1.

Make Complete Words

Use the letters in the boxes to create words.

Circle the word beginnings that will correctly form spelling words. Then write the word on the lines below.

b	gl
h	t
squ	sh
m	y
are	

g	th
q	o
e	d
u	l
ares	

st	z
w	b
p	ch
i	q
airs	

pl	cr
br	sh
ch	cl
dr	th
eir	

z	j
w	b
u	c
q	e
ear	

x	c
k	th
p	tr
m	u
ears	

_____

_____

_____

_____

_____

 At Home: Review the Word Study Steps to help your child spell new words.

Name _____

stairs	bare	dares	pairs	pears
mare	share	chairs	hare	square
bear	wear	glare	their	haircut

Pattern Power!

This week's spelling words contain the vowel sound /âr/. Write the spelling words that have these patterns.

/âr/ spelled *eir*

1. _____

/âr/ spelled *are*

2. _____

3. _____

4. _____

5. _____

6. _____

7. _____

8. _____

/âr/ spelled *ear*

9. _____

10. _____

11. _____

/âr/ spelled *air*

12. _____

13. _____

14. _____

15. _____

Name_____

stairs	bare	dares	pairs	pears
mare	share	chairs	hare	square
bear	wear	glare	their	haircut

Analogies

An analogy is a statement that compares sets of words that are alike in some way: **Night** is to **day** as **black** is to **white**. This analogy points out that **night** and **day** are opposite in the same way that **black** and **white** are opposite.

Use the spelling words to complete the analogies below.

1. **Grass** is to **trim** as **hair** is to _____.

2. **Toad** is to **frog** as **rabbit** is to _____.

3. **Ball** is to **block** as **circle** is to _____.

4. **Kitten** is to **cat** as **cub** is to _____.

Definition, Please!

Write the spelling word that matches each definition.

5. A female horse _____

6. To shine brightly _____

7. A flight of steps _____

8. Not covered _____

9. To let other people use something _____

10. Groups of two _____

There are five spelling mistakes in these riddles. Circle the misspelled words. Write the words correctly on the lines below.

Riddles

What have four legs but can't walk?   **Answer:** Chares

What did the peres say to the hayr?   **Answer:** Nothing, they can't talk!

What go up and down but never move?   **Answer:** Steirs

Why doesn't a beir wear shoes?   **Answer:** So he can go barefoot.

1. _____   4. _____

2. _____   5. _____

3. _____

Writing Activity

Write two riddles of your own. Use at least two spelling words in each riddle.

_____

_____

_____

_____

_____

_____

Name _____

Look at the words in each set below. One word in each set is spelled correctly. Look at Sample A. The letter next to the correctly spelled word in Sample A has been shaded in. Do Sample B yourself. Shade the letter of the word that is spelled correctly. When you are sure you know what to do, go on with the rest of the page.

Sample A:

Ⓐ caer
Ⓑ cere
Ⓒ care
Ⓓ carr

Sample B:

Ⓔ yeer
Ⓕ year
Ⓖ yere
Ⓗ yare

1. Ⓐ staires
   Ⓑ stayres
   Ⓒ stairs
   Ⓓ steres

2. Ⓔ mare
   Ⓕ marr
   Ⓖ maar
   Ⓗ mair

3. Ⓐ baer
   Ⓑ bear
   Ⓒ bair
   Ⓓ bere

4. Ⓔ bare
   Ⓕ bere
   Ⓖ bair
   Ⓗ baire

5. Ⓐ shair
   Ⓑ shaire
   Ⓒ shere
   Ⓓ share

6. Ⓔ wair
   Ⓕ wear
   Ⓖ waer
   Ⓗ waire

7. Ⓐ daers
   Ⓑ dars
   Ⓒ dairs
   Ⓓ dares

8. Ⓔ chares
   Ⓕ chiars
   Ⓖ cheres
   Ⓗ chairs

9. Ⓐ glar
   Ⓑ glare
   Ⓒ glere
   Ⓓ gler

10. Ⓔ piars
    Ⓕ peres
    Ⓖ pairs
    Ⓗ payrs

11. Ⓐ haire
    Ⓑ haer
    Ⓒ hare
    Ⓓ hayr

12. Ⓔ their
    Ⓕ thayr
    Ⓖ thier
    Ⓗ theire

13. Ⓐ perrs
    Ⓑ parrs
    Ⓒ pears
    Ⓓ paers

14. Ⓔ squaire
    Ⓕ square
    Ⓖ sqare
    Ⓗ squair

15. Ⓐ hayrecut
    Ⓑ herrcut
    Ⓒ heyrcut
    Ⓓ haircut

© Macmillan/McGraw-Hill

Name_____

Fold back the paper along the dotted line. Use the blanks to write each word as it is read aloud. When you finish the test, unfold the paper. Use the list at the right to correct any spelling mistakes.

1. _____
2. _____
3. _____
4. _____
5. _____
6. _____
7. _____
8. _____
9. _____
10. _____
11. _____
12. _____
13. _____
14. _____
15. _____

Review Words   16. _____
17. _____
18. _____

Challenge Words   19. _____
20. _____

1. turns
2. first
3. herds
4. learn
5. purr
6. third
7. earn
8. nurse
9. perch
10. girls
11. firm
12. word
13. world
14. serve
15. worth
16. bare
17. bear
18. stairs
19. perfect
20. Thursday

At Home: Help your child practice the words he or she missed to prepare for the Posttest.

Name_____

Using the Word Study Steps

1. LOOK at the word.

2. SAY the word aloud.

3. STUDY the letters in the word.

4. WRITE the word.

5. CHECK the word.
   Did you spell the word right?
   If not, go back to step 1.

Find and Circle

Where are the spelling words?

W	O	R	L	D	Z	P	X	Q	C
O	C	F	G	C	N	U	R	S	E
R	D	A	W	E	G	R	E	J	A
T	H	I	R	D	I	R	I	P	R
H	E	J	U	I	R	Q	S	E	N
W	R	C	F	T	L	E	A	R	N
F	D	U	I	O	S	P	D	C	D
I	S	E	R	V	E	K	L	H	F
R	A	K	S	G	T	U	R	N	S
M	A	A	T	J	W	O	R	D	R

At Home: Review the Word Study Steps to help your child
spell new words.

© Macmillan/McGraw-Hill

Name_____

turns	learn	earn	girls	world
first	purr	nurse	firm	serve
herds	third	perch	word	worth

This week's spelling words contain the vowel sound /ûr/. Write the spelling words that have these patterns.

/ûr/ spelled _ur_

1. _____

2. _____

3. _____

/ûr/ spelled _er_

4. _____

5. _____

6. _____

/ûr/ spelled _ear_

7. _____

8. _____

/ûr/ spelled _ir_

12. _____

13. _____

14. _____

15. _____

/ûr/ spelled _or_

9. _____

10. _____

11. _____

turns	learn	earn	girls	world
first	purr	nurse	firm	serve
herds	third	perch	word	worth

What's the Word?

Complete each sentence with a spelling word.

1. There are hundreds of different trees in the _____.

2. _____ of elephants travel in packs in Africa.

3. Vets _____ sick animals to health.

4. A man from Norway was the _____ person to reach the South Pole.

5. Betsy will _____ as class president this year.

6. When I pet my cat I hear her _____.

7. The bird sang all day as she sat on her _____.

8. I _____ 50 dollars a week for cleaning out the cage.

9. Jack was the _____ person in the line.

10. The truck cannot make wide _____ around the corners.

11. The _____ was very hard to spell.

12. The painting of the animals was _____ a lot of money.

Find the Opposites

Write the spelling word that is the opposite of each word.

13. boys _____

14. soft _____

15. forget _____

16. last _____

Name_____

There are six spelling mistakes in this essay. Circle the misspelled words. Write the words correctly on the lines below.

Exploring Ecosystems

A lot of people want to explore space, but I want to explore this wurld. There is so much to see and do right here on this planet. I want to lern as much as I can about the deserts and oceans here on earth. I want to study hurds of camels as they cross the Sahara. I want to observe eagles as they pirch on a mountain. It will be difficult, but it will be wurth it. I may even be the furst person to discover a new ecosystem!

1. _____   4. _____

2. _____   5. _____

3. _____   6. _____

Writing Activity

Imagine that you are an explorer. Write about a new ecosystem that you just found for the first time. Use at least four spelling words.

_____

_____

_____

_____

_____

_____

Name_____

Look at the words in each set below. One word in each set
is spelled correctly. Look at Sample A. The letter next to the
correctly spelled word in Sample A has been shaded in. Do
Sample B yourself. Shade the letter of the word that is spelled
correctly. When you are sure you know what to do, go on with
the rest of the page.

Sample A:

Ⓐ hir
Ⓑ herr
Ⓒ hur
Ⓓ her

Sample B:

Ⓔ dirt
Ⓕ dert
Ⓖ dort
Ⓗ durt

1. Ⓐ turns
   Ⓑ tirns
   Ⓒ terns
   Ⓓ terrns

2. Ⓔ forst
   Ⓕ first
   Ⓖ ferst
   Ⓗ furst

3. Ⓐ hirds
   Ⓑ heirds
   Ⓒ herds
   Ⓓ hurds

4. Ⓔ lern
   Ⓕ liern
   Ⓖ learn
   Ⓗ lirn

5. Ⓐ purr
   Ⓑ pir
   Ⓒ pur
   Ⓓ perr

6. Ⓔ thurd
   Ⓕ third
   Ⓖ therd
   Ⓗ thord

7. Ⓐ erne
   Ⓑ arne
   Ⓒ earn
   Ⓓ urne

8. Ⓔ nirse
   Ⓕ nurse
   Ⓖ nerse
   Ⓗ narse

9. Ⓐ pirch
   Ⓑ purch
   Ⓒ perch
   Ⓓ pourch

10. Ⓔ goils
    Ⓕ gerls
    Ⓖ gurls
    Ⓗ girls

11. Ⓐ furm
    Ⓑ ferm
    Ⓒ firem
    Ⓓ firm

12. Ⓔ wurd
    Ⓕ word
    Ⓖ wird
    Ⓗ werd

13. Ⓐ world
    Ⓑ wurld
    Ⓒ wirld
    Ⓓ warld

14. Ⓔ sirve
    Ⓕ sierve
    Ⓖ surve
    Ⓗ serve

15. Ⓐ wurth
    Ⓑ warth
    Ⓒ wirth
    Ⓓ worth

Fold back the paper along the dotted line. Use the blanks to write each word as it is read aloud. When you finish the test, unfold the paper. Use the list at the right to correct any spelling mistakes.

1. _____   **1.** loop

2. _____   **2.** rude

3. _____   **3.** look

4. _____   **4.** clue

5. _____   **5.** spoon

6. _____   **6.** tube

7. _____   **7.** shook

8. _____   **8.** blue

9. _____   **9.** cubes

10. _____   **10.** goose

11. _____   **11.** mules

12. _____   **12.** gloom

13. _____   **13.** true

14. _____   **14.** shoe

15. _____   **15.** stew

Review Words 16. _____   **16.** firm

17. _____   **17.** turns

18. _____   **18.** learn

Challenge Words 19. _____   **19.** classroom

20. _____   **20.** childhood

At Home: Help your child practice the words he or she missed to prepare for the Posttest.

The Jones Family Express **83**
Book I/Unit 3

Name_____

Using the Word Study Steps

1. LOOK at the word.

2. SAY the word aloud.

3. STUDY the letters in the word.

4. WRITE the word.

5. CHECK the word.
Did you spell the word right?
If not, go back to step 1.

Crossword Puzzle

Solve the crossword puzzle with spelling words that complete the sentences.

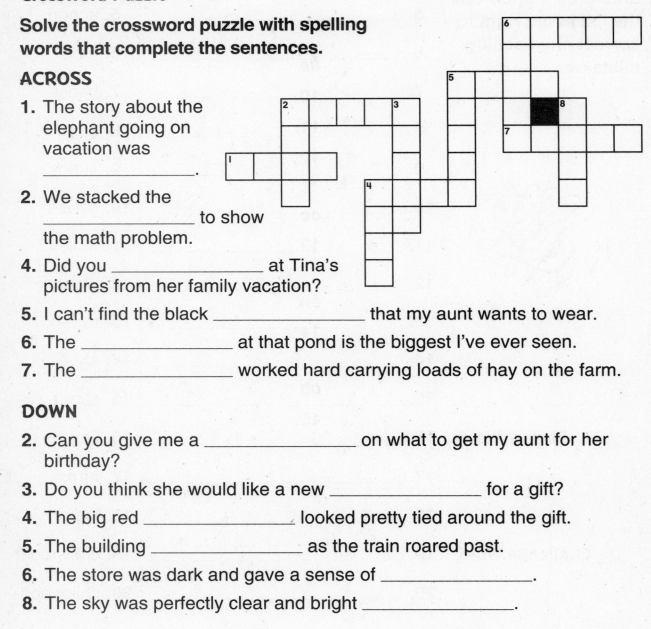

ACROSS

1. The story about the elephant going on vacation was _____.

2. We stacked the _____ to show the math problem.

4. Did you _____ at Tina's pictures from her family vacation?

5. I can't find the black _____ that my aunt wants to wear.

6. The _____ at that pond is the biggest I've ever seen.

7. The _____ worked hard carrying loads of hay on the farm.

DOWN

2. Can you give me a _____ on what to get my aunt for her birthday?

3. Do you think she would like a new _____ for a gift?

4. The big red _____ looked pretty tied around the gift.

5. The building _____ as the train roared past.

6. The store was dark and gave a sense of _____.

8. The sky was perfectly clear and bright _____.

At Home: Review the Word Study Steps to help your child spell new words.

Name_____

loop	clue	shook	goose	true
rude	spoon	blue	mules	shoe
look	tube	cubes	gloom	stew

Pattern Power!

Write the spelling words that have these patterns.

u-e

1. _____

2. _____

3. _____

4. _____

oo

5. _____

6. _____

7. _____

8. _____

9. _____

ue

10. _____

11. _____

12. _____

oe

13. _____

ew

14. _____

oo-e

15. _____

Name_____

loop	clue	shook	goose	true
rude	spoon	blue	mules	shoe
look	tube	cubes	gloom	stew

Analogies

An **analogy** is a statement that compares sets of words that are alike in some way: **Night** is to **day** as **black** is to **white**. This analogy points out that **night** and **day** are opposite in the same way that **black** and **white** are opposite.

Use spelling words to complete the analogies below.

1. **Wrong** is to **right** as **false** is to _____.

2. **Head** is to **hat** as **foot** is to _____.

3. **Ear** is to **listen** as **eye** is to _____.

4. **Neat** is to **messy** as **polite** is to _____.

In the Dictionary

Many dictionary entries have sample sentences that show how the word can be used. Complete each sample sentence with a spelling word.

5. Read the first _____ in the crossword puzzle.

6. We rode on _____ down into the Grand Canyon.

7. You can float down the river in an inner _____.

8. Make a _____ with the rope.

9. Father made a vegetable _____ for dinner.

10. Before opening the gift, she _____ the box.

11. The water will become ice _____ in the freezer.

12. I could not see anything through the fog and _____.

Name_____

There are six spelling mistakes in the travel brochure below. Circle the misspelled words. Write the words correctly on the lines below.

A family camping trip is a wonderful way to spend time together and get away from it all. You might want to louk into going in the spring. The bloo skies and green trees will be a nice change from the classroom and the office.

There is a lot to do while camping. You can go for a bike ride or take a hike on a trail. Go on a nature walk. You might see a deer or a gousse.

Cooking on a camping trip is also a lot of fun. Bring a pot, a wooden spoune, and some vegetables to make stue on the campfire. Do not forget to pack some snacks, too! It's tru that you won't have all of the comforts of home, but that makes the camping trip even more special!

1. _____   4. _____

2. _____   5. _____

3. _____   6. _____

Writing Activity

Write a postcard to a friend about a journey you have taken or would like to take. Use at least four spelling words in your description.

_____

_____

_____

_____

_____

Name _____

Look at the words in each set below. One word in each set is spelled correctly. Look at Sample A. The letter next to the correctly spelled word in Sample A has been shaded in. Do Sample B yourself. Shade the letter of the word that is spelled correctly. When you are sure you know what to do, go on with the rest of the page.

Sample A:

- Ⓐ bouk
- Ⓑ bok
- © book
- Ⓓ buuk

Sample B:

- Ⓔ too
- Ⓕ tou
- Ⓖ tue
- Ⓗ tu

1. Ⓐ lupe
   Ⓑ lewp
   © loope
   Ⓓ loop

2. Ⓔ rude
   Ⓕ roode
   Ⓖ reud
   Ⓗ rewd

3. Ⓐ louk
   Ⓑ look
   © lewk
   Ⓓ luk

4. Ⓔ cloo
   Ⓕ clewe
   Ⓖ cleu
   Ⓗ clue

5. Ⓐ spewn
   Ⓑ spoun
   © spoon
   Ⓓ spown

6. Ⓔ toob
   Ⓕ tewb
   Ⓖ teub
   Ⓗ tube

7. Ⓐ shook
   Ⓑ shuk
   © shewk
   Ⓓ shoke

8. Ⓔ bloo
   Ⓕ blue
   Ⓖ blou
   Ⓗ bluu

9. Ⓐ coobs
   Ⓑ cubes
   © coubs
   Ⓓ cewbs

10. Ⓔ gewse
    Ⓕ guse
    Ⓖ goose
    Ⓗ gouse

11. Ⓐ mules
    Ⓑ mools
    © mewles
    Ⓓ mouls

12. Ⓔ glume
    Ⓕ gloome
    Ⓖ gloom
    Ⓗ glewm

13. Ⓐ trew
    Ⓑ true
    © trou
    Ⓓ troo

14. Ⓔ shue
    Ⓕ shu
    Ⓖ shew
    Ⓗ shoe

15. Ⓐ stoow
    Ⓑ stouw
    © steuw
    Ⓓ stew

© Macmillan/McGraw-Hill

Name_____

Fold back the paper along the dotted line. Use the blanks to write each word as it is read aloud. When you finish the test, unfold the paper. Use the list at the right to correct any spelling mistakes.

1. _____ 1. coy

2. _____ 2. soil

3. _____ 3. foil

4. _____ 4. toil

5. _____ 5. coins

6. _____ 6. point

7. _____ 7. noise

8. _____ 8. loyal

9. _____ 9. boiled

10. _____ 10. spoiled

11. _____ 11. enjoys

12. _____ 12. voice

13. _____ 13. choice

14. _____ 14. soybean

15. _____ 15. joyful

Review Words 16. _____ 16. spoon

17. _____ 17. rude

18. _____ 18. shook

Challenge Words 19. _____ 19. noisy

20. _____ 20. checkpoint

At Home: Help your child practice the words he or she missed to prepare for the Posttest.

Name_____

Using the Word Study Steps

1. LOOK at the word.

2. SAY the word aloud.

3. STUDY the letters in the word.

4. WRITE the word.

5. CHECK the word.
   Did you spell the word right?
   If not, go back to step 1.

Find Rhyming Words

Circle the word in each row that rhymes with the word in dark type.

1. royal	roll	loyal	lowly
2. toil	foil	flow	fool
3. voice	choose	choice	chore
4. spoiled	older	bold	boiled
5. toys	tools	toes	enjoys
6. joins	count	coins	clowns
7. boil	soil	sore	sold
8. joint	paint	pond	point
9. boys	noise	nose	noisy
10. joy	coy	say	low

© Macmillan/McGraw-Hill

At Home: Review the Word Study Steps to help your child spell new words.

Name_____

coy	toil	noise	spoiled	choice
soil	coins	loyal	enjoys	soybean
foil	point	boiled	voice	joyful

Pattern Power
Write the spelling words with the /oi/ sound spelled:

oi

1. _____
2. _____
3. _____
4. _____
5. _____
6. _____
7. _____
8. _____

9. _____
10. _____

oy

11. _____
12. _____
13. _____
14. _____
15. _____

Syllable Power
Write the spelling words that have one syllable:

16. _____
17. _____
18. _____
19. _____
20. _____

21. _____
22. _____
23. _____
24. _____

Write the spelling words that have two syllables:

25. _____
26. _____

27. _____
28. _____

Name_____

coy	toil	noise	spoiled	choice
soil	coins	loyal	enjoys	soybean
foil	point	boiled	voice	joyful

Analogies

An analogy is a statement that compares sets of words that are alike in some way. Use spelling words to complete the analogies below.

1. **Drummer** is to **drum** as **singer** is to _____.

2. **Dark** is to **light** as **silence** is to _____.

3. **Apples** are to **fruit** as **pennies** are to _____.

4. **Bad** is to **awful** as **glad** is to _____.

5. **Head** is to **nod** as **finger** is to _____.

Define It!

Write the spelling words that have the same meaning as the words or phrases below.

6. dirt _____

7. aluminum wrap for sandwich _____

8. faithful _____

9. work hard _____

10. selection _____

11. bean used for food _____

12. shy _____

13. heated water _____

14. ruined _____

15. likes _____

Name _____

There are six spelling mistakes in the letter below. Circle the misspelled words. Write the words correctly on the lines below.

Dear Ms. Jones,

I am a loual fan of your work as an illustrator. I wanted to ask you about what it takes to do your job because I want to be an illustrator, too. Everyone says I draw well, so it seems like a good choys for me.

Your drawings make so many people joieful. I wish I could do that! Even my Uncle Bob, who never smiles, engues your drawings.

When did you decide to become an illustrator? What do you do every day? Is it hard? Are there times when you toyel over a drawing for a long time?

I think the best job for me would be one that I love. My poynt is that I think I would be a good illustrator. Thanks for being my hero!

Sincerely,
Albert Martin

1. _____   4. _____

2. _____   5. _____

3. _____   6. _____

Writing Activity

Think about a hero you have and write a letter about why you look up to him or her. Use at least three spelling words in your letter.

_____

_____

_____

_____

_____

Name_____

Look at the words in each set below. One word in each set is spelled correctly. Look at Sample A. The letter next to the correctly spelled word in Sample A has been shaded in. Do Sample B yourself. Shade the letter of the word that is spelled correctly. When you are sure you know what to do, go on with the rest of the page.

Sample A:

Ⓐ toi
Ⓑ toie
Ⓒ toye
Ⓓ toy

Sample B:

Ⓔ boy
Ⓕ boye
Ⓖ boi
Ⓗ boie

1. Ⓐ coi
   Ⓑ koie
   Ⓒ coy
   Ⓓ koy

2. Ⓔ soil
   Ⓕ soyal
   Ⓖ soyl
   Ⓗ soyll

3. Ⓐ foil
   Ⓑ foyl
   Ⓒ fiol
   Ⓓ foyal

4. Ⓔ toyal
   Ⓕ toil
   Ⓖ toyel
   Ⓗ toill

5. Ⓐ coyns
   Ⓑ coins
   Ⓒ coines
   Ⓓ cions

6. Ⓔ poynt
   Ⓕ piont
   Ⓖ poyunt
   Ⓗ point

7. Ⓐ noyse
   Ⓑ noise
   Ⓒ noize
   Ⓓ noys

8. Ⓔ loil
   Ⓕ loyel
   Ⓖ loyal
   Ⓗ loiel

9. Ⓐ boiled
   Ⓑ boyeld
   Ⓒ boyled
   Ⓓ bioled

10. Ⓔ spoyeld
    Ⓕ spoyald
    Ⓖ spoiled
    Ⓗ spoilld

11. Ⓐ enjoys
    Ⓑ enjoise
    Ⓒ enjoiys
    Ⓓ enjoice

12. Ⓔ voyce
    Ⓕ voise
    Ⓖ vouyce
    Ⓗ voice

13. Ⓐ choyce
    Ⓑ choise
    Ⓒ choyse
    Ⓓ choice

14. Ⓔ soybean
    Ⓕ soibeen
    Ⓖ soibean
    Ⓗ sueybean

15. Ⓐ joiful
    Ⓑ juoiful
    Ⓒ joyful
    Ⓓ juoyufl

Name_____

Read each sentence. If an underlined word is spelled wrong, fill in the circle that goes with that word. If no word is spelled wrong, fill in the circle below NONE. Read Sample A, then do Sample B.

A. The <u>nurse</u> told the <u>gurls</u> a funny <u>story</u>.
      A   B   C

NONE
A. Ⓐ Ⓑ Ⓒ Ⓓ

B. It is <u>harde</u> to eat <u>stew</u> without a <u>spoon</u>.
      E   F   G

NONE
B. Ⓔ Ⓕ Ⓖ Ⓗ

1. She will <u>surve</u> the <u>stew</u> <u>first</u>.
      A   B C

NONE
1. Ⓐ Ⓑ Ⓒ Ⓓ

2. There was a <u>sharp</u> <u>noys</u> in the <u>air</u>.
            E   F   G

NONE
2. Ⓔ Ⓕ Ⓖ Ⓗ

3. You <u>look</u> nice when you <u>wear</u> your <u>bloo</u> pants.
       A   B   C

NONE
3. Ⓐ Ⓑ Ⓒ Ⓓ

4. The <u>pointe</u> of the lesson was to <u>learn</u> how to <u>share</u>.
      E   F   G

NONE
4. Ⓔ Ⓕ Ⓖ Ⓗ

5. Their <u>customer</u> <u>enjoys</u> the <u>soibeen</u> bread.
     A   B   C

NONE
5. Ⓐ Ⓑ Ⓒ Ⓓ

6. The <u>girls</u> found it <u>hard</u> to listen to the <u>story</u>.
      E   F   G

NONE
6. Ⓔ Ⓕ Ⓖ Ⓗ

7. He was <u>sore</u> from climbing the <u>hard</u> <u>starres</u>.
        A   B   C

NONE
7. Ⓐ Ⓑ Ⓒ Ⓓ

8. <u>Pour</u> the <u>furst</u> batch of <u>stew</u> into the bowls.
    E   F   G

NONE
8. Ⓔ Ⓕ Ⓖ Ⓗ

9. The <u>spoiled</u> <u>gurrls</u> were <u>rude</u> to the clerk.
      A   B   C

NONE
9. Ⓐ Ⓑ Ⓒ Ⓓ

10. I will <u>shair</u> the <u>story</u> with <u>their</u> parents.
       E   F   G

NONE
10. Ⓔ Ⓕ Ⓖ Ⓗ

11. The <u>nurse</u> will <u>poore</u> the medicine into a <u>spoon</u>.
       A   B   C

NONE
11. Ⓐ Ⓑ Ⓒ Ⓓ

12. <u>Look</u> at the <u>scharp</u> <u>point</u> on the sword.
    E   F   G

NONE
12. Ⓔ Ⓕ Ⓖ Ⓗ

13. It is <u>rood</u> to make <u>noise</u> when he reads the <u>story</u>.
      A   B   C

NONE
13. Ⓐ Ⓑ Ⓒ Ⓓ

14. We painted the <u>stairs</u> <u>blue</u> to surprise the <u>nerse</u>.
                 E   F   G

NONE
14. Ⓔ Ⓕ Ⓖ Ⓗ

15. They <u>serve</u> <u>theyr</u> bread with <u>soybean</u> oil.
       A   B   C

NONE
15. Ⓐ Ⓑ Ⓒ Ⓓ

16. The <u>spoiled</u> eggs left a <u>sharp</u> smell in the <u>air</u>.
      E   F   G

NONE
16. Ⓔ Ⓕ Ⓖ Ⓗ

17. It is <u>rude</u> not to <u>share</u> your <u>stew</u>.
      A   B   C

NONE
17. Ⓐ Ⓑ Ⓒ Ⓓ

18. She <u>enjois</u> eating with the <u>blue</u> <u>spoon</u> I gave her.
      E   F   G

NONE
18. Ⓔ Ⓕ Ⓖ Ⓗ

19. <u>First</u>, we will <u>lurn</u> about <u>soybean</u> farmers.
  A   B   C

NONE
19. Ⓐ Ⓑ Ⓒ Ⓓ

20. The <u>nurse</u> took care of the <u>girls</u> who were <u>soor</u>.
      E   F   G

NONE
20. Ⓔ Ⓕ Ⓖ Ⓗ

21. He <u>spoiled</u> the surprise by taking a <u>louck</u> down the <u>stairs</u>.
    A   B   C

NONE
21. Ⓐ Ⓑ Ⓒ Ⓓ

22. The new waiter will <u>learn</u> to <u>pour</u> water and <u>serve</u> food.
                        E   F   G

NONE
22. Ⓔ Ⓕ Ⓖ Ⓗ

23. He made the <u>point</u> that you had to <u>wair</u> a <u>hard</u> hat.
             A   B   C

NONE
23. Ⓐ Ⓑ Ⓒ Ⓓ

24. Mom <u>enjoys</u> it when I <u>wear</u> my funny <u>spon</u> costume.
        E   F   G

NONE
24. Ⓔ Ⓕ Ⓖ Ⓗ

25. The loud <u>noise</u> made the <u>sore</u> cat jump in the <u>aer</u>.
        A   B   C

NONE
25. Ⓐ Ⓑ Ⓒ Ⓓ

© Macmillan/McGraw-Hill

Name_____

Fold back the paper along the dotted line. Use the blanks to write each word as it is read aloud. When you finish the test, unfold the paper. Use the list at the right to correct any spelling mistakes.

1. _____
2. _____
3. _____
4. _____
5. _____
6. _____
7. _____
8. _____
9. _____
10. _____
11. _____
12. _____
13. _____
14. _____
15. _____

Review Words 16. _____

17. _____

18. _____

Challenge Words 19. _____

20. _____

1. yawn
2. taught
3. salt
4. lawn
5. halls
6. hauls
7. hawks
8. squawk
9. bought
10. bawls
11. drawing
12. caused
13. paused
14. crawled
15. coughing
16. joyful
17. coins
18. spoiled
19. walrus
20. autumn

At Home: Help your child practice the words he or she missed to prepare for the Posttest.

Cook-a-Doodle-Do! • Book 2/Unit 4   97

© Macmillan/McGraw-Hill

Name_____

Using the Word Study Steps

1. LOOK at the word.

2. SAY the word aloud.

3. STUDY the letters in the word.

4. WRITE the word.

5. CHECK the word.
   Did you spell the word right?
   If not, go back to step 1.

X the Word

Put an X on the word in each row that does not fit the pattern.

1.	yawn	lawn	hawks	bought
2.	caused	paused	salt	hauls
3.	drawing	joy	bawls	crawled
4.	crawled	squawk	spoiled	hawks
5.	coins	taught	hauls	caused
6.	bought	salt	halls	falls
7.	drawing	lawn	broom	hawks
8.	paused	stopped	hauls	cause
9.	thought	bought	coughing	spoiled
10.	crawled	bawl	hawk	halls

At Home: Review the Word Study Steps to help your child spell new words.

Name_____

yawn	lawn	hawks	bawls	paused
taught	halls	squawk	drawing	crawled
salt	hauls	bought	caused	coughing

Pattern Power!

Write the spelling words that have these patterns.

/ô/ spelled *au*

1. _____ 2. _____ 3. _____

/ô/ spelled *aw*

4. _____ 7. _____ 9. _____

5. _____ 8. _____ 10. _____

6. _____

/ô/ spelled *a*

11. _____ 12. _____

/ô/ spelled *augh*

13. _____

/ô/ spelled *ough*

14. _____ 15. _____

Name _____

yawn	lawn	hawks	bawls	paused
taught	halls	squawk	drawing	crawled
salt	hauls	bought	caused	coughing

What's the Word?

Complete each sentence with a spelling word.

1. The chef _____ us how to bake cookies.

2. She _____ a loaf of bread at the bakery.

3. There was too much _____ in the soup.

4. Watch out! Those large _____ are trying to eat our picnic food!

5. My mother _____ when she loses her favorite recipe book.

6. We set up a lemonade stand on the front _____.

7. Margaret is _____ a picture of a strawberry shortcake.

8. She _____ the burnt cookies to the trash.

9. The smell of pepper made us start sneezing and _____.

10. After he finished dinner, Bob let out a big _____ and fell asleep.

11. You could smell the cookies all the way down our _____.

12. The smell _____ me to smile.

Synonym Alert!

For each word below, write the spelling word that has the same meaning.

13. creeped _____

14. squeal _____

15. stopped _____

Name _____

Proofreading

There are four spelling mistakes in this paragraph. Circle the misspelled words. Write the words correctly on the lines below.

Steps for making a salad:

1. Always wash your hands with soap and water before you start cooking.

2. Make sure you baught everything you need.

3. Get out the things you will need for the dressing, such as oil, vinegar, and sawlt.

4. Toss together the lettuce and the other vegetables.

5. If you have a garden next to your lown, you can add fresh vegetables to your salad.

6. Remember what you were tought for the next time you make a salad.

1. _____   3. _____

2. _____   4. _____

Writing Activity

Write the steps for another activity you like to do. Use at least three spelling words in your paragraph.

_____

_____

_____

_____

_____

_____

Name_____

Look at the words in each set below. One word in each set is spelled correctly. Look at Sample A. The letter next to the correctly spelled word in Sample A has been shaded in. Do Sample B yourself. Shade the letter of the word that is spelled correctly. When you are sure you know what to do, go on with the rest of the page.

Sample A:

Ⓐ sawlt
Ⓑ sealt
Ⓒ salt
Ⓓ sault

Sample B:

Ⓔ fawl
Ⓕ fall
Ⓖ faul
Ⓗ faol

1. Ⓐ yaun
   Ⓑ yonn
   Ⓒ yawn
   Ⓓ yann

2. Ⓔ taught
   Ⓕ tawt
   Ⓖ tawght
   Ⓗ tauht

3. Ⓐ sault
   Ⓑ sawlt
   Ⓒ selt
   Ⓓ salt

4. Ⓔ lawn
   Ⓕ laun
   Ⓖ laugn
   Ⓗ lohn

5. Ⓐ hawls
   Ⓑ haughls
   Ⓒ halls
   Ⓓ haulls

6. Ⓔ haughls
   Ⓕ hals
   Ⓖ hawls
   Ⓗ hauls

7. Ⓐ hauks
   Ⓑ hawks
   Ⓒ haks
   Ⓓ haulks

8. Ⓔ squauk
   Ⓕ squawk
   Ⓖ sqwack
   Ⓗ squock

9. Ⓔ baught
   Ⓕ bawt
   Ⓖ bought
   Ⓗ baght

10. Ⓐ boughls
    Ⓑ bals
    Ⓒ bauls
    Ⓓ bawls

11. Ⓔ drawing
    Ⓕ drauing
    Ⓖ drauwing
    Ⓗ draughing

12. Ⓐ cawsed
    Ⓑ cassed
    Ⓒ caused
    Ⓓ coused

13. Ⓐ pased
    Ⓑ poused
    Ⓒ pawsed
    Ⓓ paused

14. Ⓔ crowled
    Ⓕ crauled
    Ⓖ crawled
    Ⓗ craled

15. Ⓐ caughing
    Ⓑ coughing
    Ⓒ cawfing
    Ⓓ coghing

Name _____

Fold back the paper along the dotted line. Use the blanks to write each word as it is read aloud. When you finish the test, unfold the paper. Use the list at the right to correct any spelling mistakes.

1. _____
2. _____
3. _____
4. _____
5. _____
6. _____
7. _____
8. _____
9. _____
10. _____
11. _____
12. _____
13. _____
14. _____
15. _____

Review Words

16. _____
17. _____
18. _____

Challenge Words

19. _____
20. _____

1. found
2. town
3. shout
4. owl
5. couch
6. bow
7. scout
8. round
9. plow
10. crowd
11. proud
12. clouds
13. ground
14. louder
15. bounce
16. drawing
17. lawn
18. hauls
19. snowplow
20. outline

At Home: Help your child practice the words he or she missed to prepare for the Posttest.

Name_____

Using the Word Study Steps

1. LOOK at the word.

2. SAY the word aloud.

3. STUDY the letters in the word.

4. WRITE the word.

5. CHECK the word.
   Did you spell the word right?
   If not, go back to step 1.

Find Rhyming Words

Circle the word in each row that rhymes with the word in dark type.

1. **towel**	sole	owl	bowl
2. **ouch**	couch	foul	loud
3. **sound**	sew	down	ground
4. **out**	shout	mound	town
5. **frown**	own	town	snow
6. **found**	out	scout	round
7. **ounce**	bounce	howl	once
8. **how**	plow	plot	hope
9. **cloud**	proud	clock	draw
10. **shout**	snow	scout	stop
11. **loud**	cloud	long	couch
12. **bow**	louder	ground	plow

© Macmillan/McGraw-Hill

 At Home: Review the Word Study Steps to help your child
spell new words.

Name_____

found   owl   scout   crowd   ground
town   couch   round   proud   louder
shout   bow   plow   clouds   bounce

Pattern Power!

This week's spelling words contain the vowel sound /ou/. Write the spelling words with the /ou/ sound spelled:

ou

1. _____   6. _____

2. _____   7. _____

3. _____   8. _____

4. _____   9. _____

5. _____   10. _____

ow

11. _____   14. _____

12. _____   15. _____

13. _____

Words Within Words

Write each spelling word in which you can find the smaller word.

16. row _____   19. own _____

17. low _____   20. round _____

18. loud _____

found	owl	scout	crowd	ground
town	couch	round	proud	louder
shout	bow	plow	clouds	bounce

Analogies

An analogy is a statement that compares sets of words that are alike in some way: *Night* is to *day* as *black* is to *white*. This analogy points out that *night* and *day* are opposite in the same way that *black* and *white* are opposite.

Use the spelling words to complete the analogies below.

1. *Top* is to *bottom* as *sky* is to _____.

2. *Bad* is to *good* as *lost* is to _____.

3. *Quiet* is to *whisper* as *loud* is to _____.

4. *Flying disk* is to *throw* as *ball* is to _____.

5. *Moo* is to *cow* as *hoot* is to _____.

6. *Salute* is to *general* as _____ is to *audience*.

Define It!

Write the spelling word that matches each definition.

7. Large group of people _____

8. Big soft seat to sit on _____

9. Feeling pleased about what you did _____

10. What rain falls from _____

11. Someone who goes out to get information _____

12. Shape of a circle _____

13. A place where people live _____

14. Raised volume _____

15. Tool on a farm _____

Proofreading

There are six spelling mistakes in this paragraph. Circle the misspelled words. Write the words correctly on the lines below.

Our Class Newsletter

Our class is prowd to announce that we have come up with some new rules. The lunchroom has gotten looder over the past year. We fownd it is hard to enjoy eating our lunches. We have decided that to fix the problem we will do two things. First, everyone must sit at one of the rond tables during lunch. There must be no walking around. Second, you are not allowed to schot at each other or stomp on the growned. These new rules should make lunchtime much better. If we make the lunchroom a nicer place, everyone will want to eat there.

1. _____   4. _____

2. _____   5. _____

3. _____   6. _____

Writing Activity

Write an article for your class newsletter. Use at least three spelling words in your paragraph.

_____

_____

_____

_____

_____

Name_____

Look at the words in each set below. One word in each set is spelled correctly. Look at Sample A. The letter next to the correctly spelled word in Sample A has been shaded in. Do Sample B yourself. Shade the letter of the word that is spelled correctly. When you are sure you know what to do, go on with the rest of the page.

Sample A:

Ⓐ owt
Ⓑ ott
Ⓒ oute
Ⓓ out

Sample B:

Ⓔ brown
Ⓕ braun
Ⓖ bron
Ⓗ browne

1. Ⓐ fownd
   Ⓑ fawnd
   Ⓒ faund
   Ⓓ found

2. Ⓔ taun
   Ⓕ town
   Ⓖ tawn
   Ⓗ toun

3. Ⓐ shout
   Ⓑ showt
   Ⓒ shaut
   Ⓓ shawt

4. Ⓔ awel
   Ⓕ oal
   Ⓖ owel
   Ⓗ owl

5. Ⓐ couch
   Ⓑ cowch
   Ⓒ coch
   Ⓓ coush

6. Ⓔ bau
   Ⓕ bou
   Ⓖ bow
   Ⓗ baw

7. Ⓐ scowt
   Ⓑ scaut
   Ⓒ scout
   Ⓓ scawt

8. Ⓔ raund
   Ⓕ round
   Ⓖ rawnd
   Ⓗ rownd

9. Ⓔ plow
   Ⓕ plaugh
   Ⓖ plau
   Ⓗ plaw

10. Ⓐ craud
    Ⓑ crowd
    Ⓒ crawd
    Ⓓ crod

11. Ⓔ praud
    Ⓕ prawd
    Ⓖ proud
    Ⓗ prowd

12. Ⓐ clowds
    Ⓑ clawds
    Ⓒ clauds
    Ⓓ clouds

13. Ⓐ grownd
    Ⓑ ground
    Ⓒ graund
    Ⓓ grawnd

14. Ⓔ lowder
    Ⓕ lauwder
    Ⓖ loder
    Ⓗ louder

15. Ⓐ bownce
    Ⓑ bounse
    Ⓒ bownse
    Ⓓ bounce

Name_____

Fold back the paper along the dotted line. Use the blanks to write each word as it is read aloud. When you finish the test, unfold the paper. Use the list at the right to correct any spelling mistakes.

1. _____
2. _____
3. _____
4. _____
5. _____
6. _____
7. _____
8. _____
9. _____
10. _____
11. _____
12. _____
13. _____
14. _____
15. _____

Review Words 16. _____
17. _____
18. _____

Challenge Words 19. _____
20. _____

1. cell
2. gems
3. age
4. place
5. gyms
6. city
7. cents
8. price
9. space
10. nice
11. giant
12. changes
13. pages
14. gentle
15. message
16. crowd
17. clouds
18. found
19. giraffe
20. celebrate

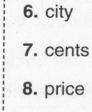 **At Home:** Help your child practice the words he or she missed to prepare for the Posttest.

Washington Weed Whackers

109

Book 2/Unit 4

Name

Using the Word Study Steps

1. LOOK at the word.

2. SAY the word aloud.

3. STUDY the letters in the word.

4. WRITE the word.

5. CHECK the word.
   Did you spell the word right?
   If not, go back to step 1.

Find and Circle

Where are the spelling words?

P	P	X	R	A	U	G	E	M	S
R	A	W	C	I	T	Y	P	L	Y
I	G	Z	K	J	D	M	L	D	M
C	E	N	T	S	G	S	A	N	E
E	S	I	V	P	E	C	C	A	S
X	M	C	Z	A	N	D	E	G	S
K	E	E	X	C	T	O	C	E	A
Q	F	T	G	E	L	B	E	N	G
C	H	A	N	G	E	S	L	O	E
G	I	A	N	T	S	W	L	R	P

At Home: Review the Word Study Steps to help your child spell new words.

© Macmillan/McGraw-Hill

Name_____

cell	place	cents	nice	pages
gems	gyms	price	giant	gentle
age	city	space	changes	message

Word Sort

Write the spelling words that have soft *c* spelled:

c

1. _____

ce

2. _____   5. _____

3. _____   6. _____

4. _____   7. _____

Write the spelling words that have soft *g* spelled:

g

8. _____   10. _____

9. _____   11. _____

ge

12. _____   14. _____

13. _____

Rhyme Time

Write the spelling word that rhymes with each word below.

15. face _____   18. pity _____

16. cages _____   19. rims _____

17. bell _____   20. ranges _____

Name_____

cell	place	cents	nice	pages
gems	gyms	price	giant	gentle
age	city	space	changes	message

It Takes Three

Write a spelling word that goes with the other two words.

1. cost, amount, _____

2. jewelry, charms, _____

3. coins, change, _____

4. town, village, _____

5. calm, tender, _____

Words in Sentences

Write a spelling word to complete each sentence.

1. There are many _____ inside a book.

2. There was a time when the Earth was a cold and frosty _____.

3. Dinosaurs lived during a different _____.

4. You get 5 _____ for each bottle you recycle.

5. We won't cut down trees to make _____ for a mall.

6. You must be _____ when you pet animals.

7. There are _____ trees in the redwood forest.

8. We had a "Save the Forest" meeting in each of the school _____.

9. My sister looked at a plant _____ under a microscope.

10. You cannot put a _____ on nature.

11. They mine for _____ in those mountains.

12. I left a _____ on her answering machine.

13. The actor _____ his costume in the play.

14. The _____ woman never litters.

15. The pollution from the _____ is hurting the forest nearby.

© Macmillan/McGraw-Hill

Name _____

Proofreading

There are six spelling mistakes in this paragraph. Circle the misspelled words. Write the words correctly on the lines below.

What You Can Do to Help Save the Planet

There are several ways to help take care of nature in your town. Write to your sitti mayor about a pollution problem. This will tell him or her that you are worried about the environment. You can raise money, too. Save your extra sence. Even spare change can help buy a tree. You can then plant it on your street or in a public park. Remember to always be jentl with animals. We all have to share the same natural space. You should be nys to any wildlife you find. Never litter. This only makes the roads and grassy areas dirty. Save your plastic bottles. Buy a ginte can to put them in. When you have enough, take them in for recycling. The most important thing you can do is this: Tell others about keeping our world clean. Pass on this important mesaje to everyone you know.

1. _____   3. _____   5. _____

2. _____   4. _____   6. _____

Writing Activity

Write a paragraph about what you can do to help the planet. Use at least three spelling words in your paragraph.

_____

_____

_____

_____

_____

Look at the words in each set below. One word in each set is spelled correctly. Look at Sample A. The letter next to the correctly spelled word in Sample A has been shaded in. Do Sample B yourself. Shade the letter of the word that is spelled correctly. When you are sure you know what to do, go on with the rest of the page.

Sample A:

Ⓐ caige
Ⓑ cage
Ⓒ caje
Ⓓ cayj

Sample B:

Ⓔ rice
Ⓕ ryse
Ⓖ ryce
Ⓗ raice

1. Ⓐ sel
   Ⓑ cell
   Ⓒ selle
   Ⓓ cel

6. Ⓔ city
   Ⓕ sity
   Ⓖ citty
   Ⓗ cety

11. Ⓔ jiant
    Ⓕ jyant
    Ⓖ giant
    Ⓗ gyant

2. Ⓔ jems
   Ⓕ jehms
   Ⓖ gems
   Ⓗ gehms

7. Ⓐ zents
   Ⓑ sence
   Ⓒ cents
   Ⓓ cense

12. Ⓐ chanjes
    Ⓑ chaynjes
    Ⓒ chainges
    Ⓓ changes

3. Ⓐ aig
   Ⓑ aje
   Ⓒ age
   Ⓓ adje

8. Ⓔ pryce
   Ⓕ price
   Ⓖ prise
   Ⓗ pryse

13. Ⓐ pages
    Ⓑ pajes
    Ⓒ payges
    Ⓓ paiges

4. Ⓔ place
   Ⓕ playce
   Ⓖ plase
   Ⓗ plaise

9. Ⓔ spase
   Ⓕ spaice
   Ⓖ spayce
   Ⓗ space

14. Ⓔ jentle
    Ⓕ gentle
    Ⓖ gentel
    Ⓗ jentel

5. Ⓐ jyms
   Ⓑ gyms
   Ⓒ jims
   Ⓓ gims

10. Ⓐ nyce
    Ⓑ nise
    Ⓒ naice
    Ⓓ nice

15. Ⓐ mecage
    Ⓑ mesadge
    Ⓒ messej
    Ⓓ message

© Macmillan/McGraw-Hill

Name_____

Fold back the paper along the dotted line. Use the blanks to write each word as it is read aloud. When you finish the test, unfold the paper. Use the list at the right to correct any spelling mistakes.

1. _____
2. _____
3. _____
4. _____
5. _____
6. _____
7. _____
8. _____
9. _____
10. _____
11. _____
12. _____
13. _____
14. _____
15. _____

Review Words 16. _____

17. _____

18. _____

Challenge Words 19. _____

20. _____

1. sale
2. sail
3. beet
4. beat
5. rode
6. road
7. rowed
8. its
9. it's
10. your
11. you're
12. there
13. they're
14. peace
15. piece
16. city
17. gems
18. space
19. seen
20. scene

© Macmillan/McGraw-Hill

At Home: Help your child practice the words he or she missed to prepare for the Posttest.

Here's My Dollar • **Book 2/Unit 4** 115

Name_____

Using the Word Study Steps

1. LOOK at the word.
2. SAY the word aloud.
3. STUDY the letters in the word.
4. WRITE the word.
5. CHECK the word.
   Did you spell the word right?
   If not, go back to step 1.

Crossword Puzzle

Write the spelling word that best matches each clue. Put the spelling word in the box that starts with the same number.

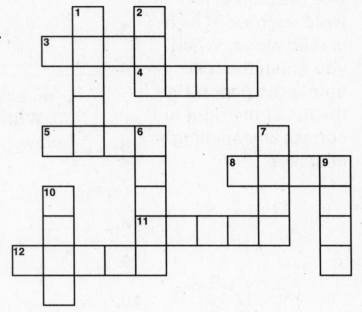

ACROSS

3. I could hear the _____ of the drum.

4. Harry started a bake _____ to raise money.

5. The baboon's face turned as red as a _____.

8. Keep _____ zoo ticket in case you need it later.

11. We _____ from one end of the zoo's lake to the other.

12. I sewed a small _____ of the quilt we made for my grandmother.

DOWN

1. We wrote a newsletter to stop wars and talk about _____.

2. The gorilla put _____ hand against the glass wall.

6. They are walking over _____ by the aquarium.

7. We wanted to fix the holes in the _____.

9. Every time I _____ my bike, I wore my helmet.

10. The sailor raised the _____ and then the boat left the dock.

At Home: Review the Word Study Steps to help your child spell new words.

Name_____

sale	beat	rowed	your	they're
sail	rode	its	you're	peace
beet	road	it's	there	piece

Homophones are words that sound alike but have different spellings and different meanings. Write the spelling words that are homophones of the words below.

1. there _____

2. peace _____

3. sale _____

4. beet _____

5. rowed _____ _____

6. it's _____

7. your _____

Which spelling words are contractions?

8. _____   9. _____   10. _____

Write the spelling words that have the sounds below.

long e

11. _____   12. _____

13. _____   14. _____

long a

15. _____   16. _____

long o

17. _____   18. _____   19. _____

© Macmillan/McGraw-Hill

Name_____

sale	beat	rowed	your	they're
sail	rode	its	you're	peace
beet	road	it's	there	piece

Homophones are words that sound alike but have different spellings and different meanings. In each sentence below, a homophone is used incorrectly. Circle the incorrect homophone and write the correct homophone on the line following the sentence.

1. Its important to help people that need you. _____

2. All of the cakes and pies were on sail for a good cause. _____

3. Is that you're mother on TV? _____

4. The rowed was long and dark. _____

5. Someday there will be piece on earth. _____

6. My family used a beat and a carrot from our garden to make soup.
_____

7. At camp we learned how to sale a boat. _____

8. We were allowed to feed the chimp a peace of banana. _____

9. The gorilla scratched it's head. _____

10. There going to open a new community center in our town.
_____

11. The group road through the jungle in a truck. _____

12. We beet last year's record by raising even more money for the zoo.
_____

13. They're are many volunteers who help at the soup kitchen.
_____

14. Your so good at listening to others. _____

15. We all road the boat to shore. _____

Name _____

Proofreading

There are seven spelling mistakes in this paragraph. Circle the misspelled words. Write the words correctly on the lines below.

Volunteer Fair

Yo're invited to our annual volunteer fair. We will have lots of ideas about how to help your community. You could adopt a raod. You could raise money for an animal shelter by holding a bake sael. You could even help build a pease of the new community center. Thire are so many ideas, you won't know where to start. Its going to be quite a fair. So please join us this Friday. Yore community needs you.

1. _____   5. _____

2. _____   6. _____

3. _____   7. _____

4. _____

Writing Activity

Write ideas you have for helping your community. Use at least three spelling words in your paragraph.

_____

_____

_____

_____

_____

_____

© Macmillan/McGraw-Hill

Name _____

Look at the words in each set below. One word in each set
is spelled correctly. Look at Sample A. The letter next to the
correctly spelled word in Sample A has been shaded in. Do
Sample B yourself. Shade the letter of the word that is spelled
correctly. When you are sure you know what to do, go on with
the rest of the page.

Sample A:

- Ⓐ soe
- Ⓑ sowe
- Ⓒ soh
- Ⓓ so

Sample B:

- Ⓔ soe
- Ⓕ sew
- Ⓖ sowe
- Ⓗ soh

1. Ⓐ sale
   Ⓑ sayle
   Ⓒ saile
   Ⓓ sayel

2. Ⓔ sayle
   Ⓕ sail
   Ⓖ cayle
   Ⓗ sayel

3. Ⓐ beet
   Ⓑ beete
   Ⓒ biet
   Ⓓ beit

4. Ⓔ beete
   Ⓕ beat
   Ⓖ beit
   Ⓗ biet

5. Ⓐ wroad
   Ⓑ raud
   Ⓒ rowd
   Ⓓ rode

6. Ⓔ rowd
   Ⓕ roud
   Ⓖ road
   Ⓗ raud

7. Ⓐ roed
   Ⓑ wrowd
   Ⓒ rowed
   Ⓓ roud

8. Ⓔ ets
   Ⓕ i'ts
   Ⓖ its
   Ⓗ itz

9. Ⓔ itz
   Ⓕ i'ts
   Ⓖ ets
   Ⓗ it's

10. Ⓐ your
    Ⓑ yure
    Ⓒ your'e
    Ⓓ yowr

11. Ⓔ yure
    Ⓕ you're
    Ⓖ your'e
    Ⓗ yowr

12. Ⓐ thier
    Ⓑ thare
    Ⓒ there
    Ⓓ theyr'e

13. Ⓐ thier
    Ⓑ theyr'e
    Ⓒ thare
    Ⓓ they're

14. Ⓔ peace
    Ⓕ peise
    Ⓖ peice
    Ⓗ pease

15. Ⓐ peice
    Ⓑ peise
    Ⓒ piece
    Ⓓ pease

Name_____

Fold back the paper along the dotted line. Use the blanks to write each word as it is read aloud. When you finish the test, unfold the paper. Use the list at the right to correct any spelling mistakes.

1. _____
2. _____
3. _____
4. _____
5. _____
6. _____
7. _____
8. _____
9. _____
10. _____
11. _____
12. _____
13. _____
14. _____
15. _____

Review Words 16. _____
17. _____
18. _____

Challenge Words 19. _____
20. _____

1. years
2. twins
3. trays
4. states
5. ashes
6. foxes
7. inches
8. flies
9. cities
10. ponies
11. bunches
12. alleys
13. lunches
14. cherries
15. daisies
16. sale
17. rode
18. you're
19. heroes
20. libraries

At Home: Help your child practice the words he or she missed to prepare for the Posttest.

My Very Own Room • Book 2/Unit 4 121

© Macmillan/McGraw-Hill

Name_____

Using the Word Study Steps

1. LOOK at the word.

2. SAY the word aloud.

3. STUDY the letters in the word.

4. WRITE the word.

5. CHECK the word.
   Did you spell the word right?
   If not, go back to step 1.

X the Word

Put an X on the word in each row that does not fit the pattern.

1.	years	lunches	ash	cherries
2.	cherry	tray	pony	bunches
3.	city	daisies	flies	states
4.	inch	lunches	bunches	cities
5.	state	ponies	ashes	trays
6.	trays	twin	cherries	alleys
7.	fox	ashes	city	munch
8.	daisies	inches	years	fly
9.	twins	foxes	alley	ponies
10.	tray	year	daisy	states
11.	rode	daisies	cherries	ponies
12.	gems	twins	years	space
13.	inches	gems	foxes	boxes
14.	years	flies	cities	ponies

At Home: Review the Word Study Steps to help your child spell new words.

Name_____

years	states	inches	ponies	lunches
twins	ashes	flies	bunches	cherries
trays	foxes	cities	alleys	daisies

This week's spelling list contains plural words. Plurals are words that name more than one thing.

Write the spelling words for each of these plural endings.

s

1. _____

2. _____

3. _____

4. _____

5. _____

es

11. _____

12. _____

13. _____

14. _____

15. _____

y to *i* + *-es*

6. _____

7. _____

8. _____

9. _____

10. _____

Find the Base Word

Write the base word of each plural noun.

16. flies _____

17. ponies _____

18. bunches _____

Name_____

years	states	inches	ponies	lunches
twins	ashes	flies	bunches	cherries
trays	foxes	cities	alleys	daisies

Part of the Group

Add the spelling word that belongs in each group below.

Fruits

1. apples, grapes, _____

Baby animals

2. calves, kittens, _____

Places to live

3. towns, villages, _____

Units of time

4. days, months, _____

Animals

5. sheep, bears, _____

Flowers

6. roses, lilies, _____

A Clue for You

7. They are small streets behind buildings. _____

8. You carry food on them. _____

9. There are 50 of these in the United States. _____

10. They buzz through the air. _____

11. What is left after something burns. _____

12. What students bring to school to eat. _____

13. Groups of something. _____

14. There are 12 of these in 1 foot. _____

15. Two people who look exactly alike. _____

Name _____

Proofreading

There are six spelling mistakes in this paragraph. Circle the misspelled words. Write the words correctly on the lines below.

Melody and Melissa were tooins, but they couldn't have been more different. Melody loved picking daisys, arranging flowers, and playing with her stuffed poonys. Melissa loved flis and insects and crawling around in the dirt. The problem was that they shared a room. Melody liked the room to be neat with boonchs of flowers in all the windows. Melissa was far from neat. She tracked in mud and brought bugs into the room. It had been a problem for many years. One day Melody decided that maybe she and Melissa should divide the room in two. That way they could both get what they wanted. Melissa thought it was a great idea. They hung a white sheet a few inshs from the ceiling. Now Melody's room is always beautiful, and Melissa's room is always messy. They are the happiest sisters around.

1. _____

2. _____

3. _____

4. _____

5. _____

6. _____

Writing Activity

If you could have your dream room, what would it be like? Use at least three spelling words in your paragraph.

_____

_____

_____

_____

_____

Name_____

Look at the words in each set below. One word in each set is spelled correctly. Look at Sample A. The letter next to the correctly spelled word in Sample A has been shaded in. Do Sample B yourself. Shade the letter of the word that is spelled correctly. When you are sure you know what to do, go on with the rest of the page.

Sample A:

Ⓐ keys
Ⓑ keeze
Ⓒ keyes
Ⓓ keies

Sample B:

Ⓔ ladys
Ⓕ ladees
Ⓖ laides
Ⓗ ladies

1. Ⓐ yeares
   Ⓑ yiers
   Ⓒ years
   Ⓓ yeirs

2. Ⓔ twins
   Ⓕ twinz
   Ⓖ twiness
   Ⓗ twyns

3. Ⓐ traies
   Ⓑ trays
   Ⓒ trayies
   Ⓓ traes

4. Ⓔ staties
   Ⓕ statez
   Ⓖ states
   Ⓗ statses

5. Ⓐ ashs
   Ⓑ ashies
   Ⓒ ashez
   Ⓓ ashes

6. Ⓔ foxes
   Ⓕ foxies
   Ⓖ foxs
   Ⓗ foxses

7. Ⓐ inchs
   Ⓑ inchies
   Ⓒ inchez
   Ⓓ inches

8. Ⓔ flys
   Ⓕ flyes
   Ⓖ flies
   Ⓗ fliez

9. Ⓔ cityies
   Ⓕ cities
   Ⓖ citys
   Ⓗ citees

10. Ⓐ poneis
    Ⓑ poneese
    Ⓒ poneez
    Ⓓ ponies

11. Ⓔ bunches
    Ⓕ bunchs
    Ⓖ bunschs
    Ⓗ bunchez

12. Ⓐ alleyies
    Ⓑ alleyes
    Ⓒ alleys
    Ⓓ alleies

13. Ⓐ lunchs
    Ⓑ lunches
    Ⓒ lunschs
    Ⓓ lunchez

14. Ⓔ cherrys
    Ⓕ cherryis
    Ⓖ cherryies
    Ⓗ cherries

15. Ⓐ daisys
    Ⓑ daisees
    Ⓒ daisies
    Ⓓ daysies

© Macmillan/McGraw-Hill

Name

Read each sentence. If an underlined word is spelled wrong, fill in the circle that goes with that word. If no word is spelled wrong, fill in the circle below NONE. Read Sample A, and do Sample B.

A. The <u>giant</u> slipped on a <u>pice</u> of <u>ice</u>.
        A   B   C

NONE
A. Ⓐ ⬤Ⓑ Ⓒ Ⓓ

B. There was a <u>shout</u> from the <u>crowd</u> when the team <u>loost</u>.
           E   F   G

NONE
B. Ⓔ Ⓕ Ⓖ Ⓗ

1. The <u>croud</u> got <u>louder</u> as the hero came down the <u>road</u>.
     A   B   C

NONE
1. Ⓐ Ⓑ Ⓒ Ⓓ

2. I <u>water</u> my <u>daisys</u> with the pitcher you <u>bought</u> me.
    E   F   G

NONE
2. Ⓔ Ⓕ Ⓖ Ⓗ

3. After my dad <u>flies</u> to a <u>city</u>, he <u>wawks</u> to his hotel.
          A   B   C

NONE
3. Ⓐ Ⓑ Ⓒ Ⓓ

4. The <u>town</u> did not believe that the <u>giyant</u> was <u>nice</u>.
      E   F   G

NONE
4. Ⓔ Ⓕ Ⓖ Ⓗ

5. Your <u>piece</u> of cake is two <u>inches</u> bigger than mine.
  A B   C

NONE
5. Ⓐ Ⓑ Ⓒ Ⓓ

6. <u>They're</u> going to leave you a <u>messige</u> if they get <u>lost</u>.
   E   F   G

NONE
6. Ⓔ Ⓕ Ⓖ Ⓗ

7. Bob <u>bete</u> Mark in a race <u>that</u> <u>night</u>.
     A   B C

NONE
7. Ⓐ Ⓑ Ⓒ Ⓓ

8. <u>Yorre</u> uncle lived in ten different <u>states</u> in ten <u>years</u>.
  E   F   G

NONE
8. Ⓔ Ⓕ Ⓖ Ⓗ

9. The vase of <u>daisies</u> looked <u>nice</u> on the <u>rownd</u> table.
         A   B   C

NONE
9. Ⓐ Ⓑ Ⓒ Ⓓ

10. Gather <u>around</u> and I will tell you about the <u>giant</u> who
       E   F

scared the <u>toun</u>.
       G

NONE
10. Ⓔ Ⓕ Ⓖ Ⓗ

11. It is hard to understand <u>your</u> <u>message</u> when you <u>showt</u>
                       A   B   C

into the phone.

NONE
11. Ⓐ Ⓑ Ⓒ Ⓓ

12. It would not be good if the three <u>inchs</u> of <u>water</u> in the
                               E   F

basement turned to <u>ice</u>.
               G

NONE
12. Ⓔ Ⓕ Ⓖ Ⓗ

© Macmillan/McGraw-Hill

13. Cara <u>bought</u> a <u>piece</u> of pie at the <u>city</u> fair.
          A   B   C

            NONE
13. Ⓐ Ⓑ Ⓒ Ⓓ

14. <u>They're</u> being <u>lowder</u> than a whole <u>crowd</u>.
       E   F   G

            NONE
14. Ⓔ Ⓕ Ⓖ Ⓗ

15. I heard my mother <u>shout</u> when she <u>fell</u> on the <u>ise</u>.
                    A   B   C

            NONE
15. Ⓐ Ⓑ Ⓒ Ⓓ

16. On one of her <u>walks</u>, she <u>lost</u> the necklace he gave her
                 E   F

    many <u>yeres</u> ago.
         G

            NONE
16. Ⓔ Ⓕ Ⓖ Ⓗ

17. I <u>bawt</u> my mother a suitcase to use when she <u>flies</u> to
      A   B

    different <u>states</u> for work.
            C

            NONE
17. Ⓐ Ⓑ Ⓒ Ⓓ

18. For <u>years</u> that team <u>beat</u> every other team they played
         E   F

    on the <u>rood</u>.
          G

            NONE
18. Ⓔ Ⓕ Ⓖ Ⓗ

19. In his <u>message</u> he told us he slipped on a <u>peese</u> of <u>ice</u>.
           A   B   C

            NONE
19. Ⓐ Ⓑ Ⓒ Ⓓ

20. The <u>town</u> came together to <u>beat</u> the <u>giant</u> at his own game.
        E   F   G

            NONE
20. Ⓔ Ⓕ Ⓖ Ⓗ

21. The <u>crowd</u> got <u>louder</u> as Bill swam through the <u>wauter</u>.
        A   B   C

            NONE
21. Ⓐ Ⓑ Ⓒ Ⓓ

22. The <u>nise</u> man who <u>walks</u> by our house wears a <u>round</u> hat.
       E   F   G

            NONE
22. Ⓔ Ⓕ Ⓖ Ⓗ

23. <u>They're</u> visiting the <u>sittee</u> down the <u>road</u> today.
      A   B   C

            NONE
23. Ⓐ Ⓑ Ⓒ Ⓓ

24. I visited so many <u>staits</u> this <u>summer</u> that I don't know
                E   F

    where I <u>lost</u> my jacket.
           G

            NONE
24. Ⓔ Ⓕ Ⓖ Ⓗ

25. When she <u>flys</u> to see her grandmother, she brings
           A

    <u>daisies</u> tied with a few <u>inches</u> of ribbon.
      B   C

            NONE
25. Ⓐ Ⓑ Ⓒ Ⓓ

© Macmillan/McGraw-Hill

Name_____

Fold back the paper along the dotted line. Use the blanks to write each word as it is read aloud. When you finish the test, unfold the paper. Use the list at the right to correct any spelling mistakes.

1. _____

2. _____

3. _____

4. _____

5. _____

6. _____

7. _____

8. _____

9. _____

10. _____

11. _____

12. _____

13. _____

14. _____

15. _____

Review Words 16. _____

17. _____

18. _____

Challenge Words 19. _____

20. _____

1. airplane

2. daytime

3. birthday

4. daylight

5. hairdo

6. notebook

7. birdhouse

8. barefoot

9. headlight

10. sometime

11. someone

12. newspaper

13. sidewalks

14. basketball

15. stagecoach

16. states

17. inches

18. cities

19. somebody

20. handwriting

© Macmillan/McGraw-Hill

At Home: Help your child practice the words he or she missed to prepare for the Posttest.

Boom Town • **Book 2/Unit 5** 129

Name_____

Using the Word Study Steps

1. LOOK at the word.
2. SAY the word aloud.
3. STUDY the letters in the word.

4. WRITE the word.
5. CHECK the word.
   Did you spell the word right?
   If not, go back to step 1.

Compound Riddles

Join two words from the riddle to make a compound word.

1. A foot that is bare _____

2. The light of the day _____

3. Paper where you read the news _____

4. The day of your birth _____

5. The house of a bird _____

6. The way someone will do your hair _____

7. Time during the day _____

8. A ball you shoot in a basket _____

9. A book in which you write a note _____

10. Place for walks on the side of the road _____

Make a Compound Word

A compound word is made up of two or more smaller words.
Draw lines connecting the words that form other compound words.

11. some   coach

12. air   one

13. head   time

14. some   plane

15. stage   light

© Macmillan/McGraw-Hill

At Home: Review the Word Study Steps to help your child
spell new words.

Name_____

airplane	daylight	birdhouse	sometime	sidewalks
daytime	hairdo	barefoot	someone	basketball
birthday	notebook	headlight	newspaper	stagecoach

What's in a Word?

Compound words are made up of smaller words. Write the spelling words that have the following words in them.

1. some _____ _____

2. light _____ _____

3. time _____ _____

4. day _____ _____ _____

Order Please!

Write each group of spelling words in alphabetical order.

birdhouse, airplane, basketball, barefoot, birthday

5. _____

6. _____

7. _____

8. _____

9. _____

notebook, hairdo, newspaper, headlight, daytime

10. _____

11. _____

12. _____

13. _____

14. _____

someone, sometime, sidewalks, stagecoach

15. _____

16. _____

17. _____

18. _____

Name_____

airplane   daylight   birdhouse   sometime   sidewalks
daytime   hairdo   barefoot   someone   basketball
birthday   notebook   headlight   newspaper   stagecoach

What's the Word?

Complete each sentence with a spelling word.

1. My brother delivers the _____ on our street to make money.

2. I wish _____ would hire me.

3. Put on your shoes. You can't go to work _____.

4. I often wake up before _____.

5. The woman went to the beauty salon for a new _____.

6. Write down your ideas in a _____ so you do not forget them.

7. The car's _____ was not working.

8. The _____ along the streets were full of people walking.

9. The store sells shoes for kids who play soccer, baseball, and

   _____.

10. She works at night because she is too busy during the _____.

11. My _____ is a day that I stay home from work and relax.

12. I would like to open my own business _____ in the future.

Where Would It Be?

**Write the spelling word that would most likely be found in each
of these places.**

13. in the backyard _____

14. at an airport _____

15. in a movie about the old West _____

Name_____

**There are six spelling mistakes in the letter. Circle the
misspelled words. Write the words correctly on the lines below.**

Dear Mr. Taylor,

I am writing to you for some advice on a business idea I had. I read about
your sports store in the nuespapper. You seem like sumone who could help me.

I had the idea one day while I was walking home. My feet were hurting
because I had been playing basckettebal bearfoot. My idea was to create a
mailing list of customers. With each name, you could also list the person's
burthdea. Your store could use the list to send letters reminding your
customers to buy a new pair of shoes. You could even include a coupon!
Maybe you could sell the list to different stores in town. The barber shop
could send a coupon for a haredoe.

What do you think? If you are interested, please write to me.

Regards,
Josh Curtain

1. _____   4. _____

2. _____   5. _____

3. _____   6. _____

Writing Activity

**Write a paragraph describing a business you would like to start.
Use at least four spelling words in your description.**

_____

_____

_____

_____

_____

Name_____

Look at the words in each set below. One word in each set is spelled correctly. Look at Sample A. The letter next to the correctly spelled word in Sample A has been shaded in. Do Sample B yourself. Shade the letter of the word that is spelled correctly. When you are sure you know what to do, go on with the rest of the page.

Sample A:

Ⓐ whithout
Ⓑ wittout
Ⓒ without
Ⓓ wethout

Sample B:

Ⓔ owtside
Ⓕ outsighted
Ⓖ ootside
Ⓗ outside

1. Ⓐ areplain
   Ⓑ airplane
   Ⓒ airplain
   Ⓓ errplane

2. Ⓔ daytime
   Ⓕ daytyme
   Ⓖ daitime
   Ⓗ daytim

3. Ⓐ birthdai
   Ⓑ birtday
   Ⓒ birthday
   Ⓓ berthday

4. Ⓔ daylite
   Ⓕ daelyte
   Ⓖ daylight
   Ⓗ deylight

5. Ⓐ hairdo
   Ⓑ hayredo
   Ⓒ haredo
   Ⓓ herrdo

6. Ⓔ knotebook
   Ⓕ nootbook
   Ⓖ notebook
   Ⓗ notbook

7. Ⓐ birdhause
   Ⓑ birdhaus
   Ⓒ birdhous
   Ⓓ birdhouse

8. Ⓔ barefut
   Ⓕ barefoot
   Ⓖ baerfoot
   Ⓗ bairfoot

9. Ⓔ headlite
   Ⓕ hedlight
   Ⓖ headlyte
   Ⓗ headlight

10. Ⓐ soumtime
    Ⓑ sometime
    Ⓒ sumtime
    Ⓓ sometyme

11. Ⓔ someone
    Ⓕ sumwon
    Ⓖ somewon
    Ⓗ somewan

12. Ⓐ newspaper
    Ⓑ newpaper
    Ⓒ newspapper
    Ⓓ knewspaper

13. Ⓐ sidewoks
    Ⓑ sidewalks
    Ⓒ sydewalkes
    Ⓓ sidewaulks

14. Ⓔ basketbull
    Ⓕ basketbal
    Ⓖ basketbol
    Ⓗ basketball

15. Ⓐ stagcoche
    Ⓑ stajcoach
    Ⓒ stagecoah
    Ⓓ stagecoach

Name_____

Fold back the paper along the dotted line. Use the blanks to write each word as it is read aloud. When you finish the test, unfold the paper. Use the list at the right to correct any spelling mistakes.

1. _____
2. _____
3. _____
4. _____
5. _____
6. _____
7. _____
8. _____
9. _____
10. _____
11. _____
12. _____
13. _____
14. _____
15. _____

Review Words 16. _____
17. _____
18. _____

Challenge Words 19. _____
20. _____

1. names
2. named
3. naming
4. hopes
5. hoped
6. hoping
7. dances
8. danced
9. dancing
10. drops
11. dropped
12. dropping
13. wraps
14. wrapped
15. wrapping
16. airplane
17. someone
18. newspaper
19. driving
20. traded

At Home: Help your child practice the words he or she missed to prepare for the Posttest.

Beatrice's Goat • Book 2/Unit 5 135

Name_____

Using the Word Study Steps

1. LOOK at the word.

2. SAY the word aloud.

3. STUDY the letters in the word.

4. WRITE the word.

5. CHECK the word.
   Did you spell the word right?
   If not, go back to step 1.

X the Word

Put an X on the word in each row that does not fit the pattern.

1. named	jumped	grabbed	stirs
2. wrapped	wraps	lined	hurried
3. lives	hopes	gives	giving
4. drop	dropping	playing	wrapping
5. dances	dancing	hoping	running
6. naming	shake	shaking	splitting
7. dropped	wrapped	hope	hoped
8. wraps	names	dances	play
9. digging	forgets	naming	losing
10. wrap	hoped	stopped	mopped

At Home: Review the Word Study Steps to help your child spell new words.

Name_____

names	hopes	dances	drops	wraps
named	hoped	danced	dropped	wrapped
naming	hoping	dancing	dropping	wrapping

Pattern Power!

Write the spelling words that show what you do before adding *-ed* or *-ing*.

drop *e* and add *-ed*

1. _____
2. _____
3. _____

drop *e* and add *-ing*

6. _____
7. _____
8. _____

double final consonant and add *-ed*

4. _____
5. _____

double final consonant and add *-ing*

9. _____
10. _____

Rhyme Time

Write a spelling word that rhymes with each of these words.

11. stops _____
12. ropes _____
13. maps _____
14. games _____
15. chances _____

Name_____

names	hopes	dances	drops	wraps
named	hoped	danced	dropped	wrapped
naming	hoping	dancing	dropping	wrapping

What's the Word?

Complete each sentence with a spelling word.

1. She was _____ woman of the year for her good work.

2. The people _____ and sang for joy.

3. The mother was so happy that a tear _____ from her eye.

4. The child _____ to go to college in the future.

5. I _____ to spend a year helping others.

6. My brother _____ up sandwiches to give to the homeless.

7. The girl sings and _____ to get parts in musicals.

8. The group _____ one winner of the service award each week.

9. When Dad _____ you off at school, go right inside.

10. We _____ the food up for the soup kitchen.

Find the Base Words

Write the base word of each -ing word.

11. naming _____

12. wrapping _____

13. dancing _____

14. dropping _____

15. hoping _____

Name_____

There are six spelling mistakes in the speech below. Circle the misspelled words. Write the words correctly on the lines below.

Welcome, students, parents, and teachers, to this assembly!

This year our school is giving an award to the student who has done the most to help others. One student has shown us that it does not matter what you are naimmed or where you live. Everyone can find ways to help others.

She hoopes to be a role model for other students. Her actions prove that even small things can make a difference. Our winner spent time droping food off at a soup kitchen with her family and wraping small gifts for people in a nursing home. She has also shared her talents with others, danceing in performances at a local hospital.

For all these reasons and more, we are nameing Susan Harper our student of the year!

1. _____   4. _____

2. _____   5. _____

3. _____   6. _____

Writing Activity

Write a paragraph about how you could use your talents to help others. Use at least four spelling words in your description.

_____

_____

_____

_____

_____

_____

Name_____

Look at the words in each set below. One word in each set
is spelled correctly. Look at Sample A. The letter next to the
correctly spelled word in Sample A has been shaded in. Do
Sample B yourself. Shade the letter of the word that is spelled
correctly. When you are sure you know what to do, go on with
the rest of the page.

Sample A:

Ⓐ skips
Ⓑ skipse
Ⓒ skeps
Ⓓ skipce

Sample B:

Ⓔ hoppet
Ⓕ hopped
Ⓖ hoppt
Ⓗ haupped

1. Ⓐ naimes
   Ⓑ naimz
   Ⓒ namses
   Ⓓ names

2. Ⓔ named
   Ⓕ naimed
   Ⓖ naimd
   Ⓗ naymd

3. Ⓐ namin
   Ⓑ naiming
   Ⓒ nameing
   Ⓓ naming

4. Ⓔ haups
   Ⓕ haupes
   Ⓖ hopes
   Ⓗ hopps

5. Ⓐ haupt
   Ⓑ hoped
   Ⓒ hauped
   Ⓓ howpt

6. Ⓔ hauping
   Ⓕ hopin
   Ⓖ hoping
   Ⓗ hoppin

7. Ⓐ dances
   Ⓑ danses
   Ⓒ dancis
   Ⓓ dansis

8. Ⓔ dansed
   Ⓕ dancd
   Ⓖ danst
   Ⓗ danced

9. Ⓔ dancing
   Ⓕ dansing
   Ⓖ dancign
   Ⓗ dancin

10. Ⓐ drawps
    Ⓑ drops
    Ⓒ draups
    Ⓓ dropes

11. Ⓔ draupt
    Ⓕ drawpt
    Ⓖ dropped
    Ⓗ droped

12. Ⓐ draupin
    Ⓑ dropping
    Ⓒ droppin
    Ⓓ droppinge

13. Ⓐ rapse
    Ⓑ wrapse
    Ⓒ rapps
    Ⓓ wraps

14. Ⓔ rappt
    Ⓕ wrappt
    Ⓖ wrapt
    Ⓗ wrapped

15. Ⓐ wrapping
    Ⓑ rappin
    Ⓒ wrappin
    Ⓓ wrappen

Name _____

Fold back the paper along the dotted line. Use the blanks to write each word as it is read aloud. When you finish the test, unfold the paper. Use the list at the right to correct any spelling mistakes.

1. _____   **1.** tries

2. _____   **2.** tried

3. _____   **3.** trying

4. _____   **4.** dries

5. _____   **5.** dried

6. _____   **6.** drying

7. _____   **7.** hurries

8. _____   **8.** hurried

9. _____   **9.** hurrying

10. _____   **10.** studies

11. _____   **11.** studied

12. _____   **12.** studying

13. _____   **13.** plays

14. _____   **14.** played

15. _____   **15.** playing

Review Words   16. _____   **16.** dances

17. _____   **17.** hoping

18. _____   **18.** wrapping

Challenge Words   19. _____   **19.** obeyed

20. _____   **20.** worrying

© Macmillan/McGraw-Hill

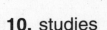 **At Home:** Help your child practice the words he or she missed to prepare for the Posttest.

Name _____

Using the Word Study Steps

1. LOOK at the word.

2. SAY the word aloud.

3. STUDY the letters in the word.

4. WRITE the word.

5. CHECK the word.
   Did you spell the word right?
   If not, go back to step 1.

Find and Circle

Where are the spelling words?

T	R	Y	I	N	G	E	H	B	V	X	M
X	E	R	Q	P	Y	T	U	J	H	H	H
S	T	U	D	I	E	D	R	Z	N	U	J
T	R	I	E	S	R	R	R	U	E	R	F
P	I	V	Q	U	Y	I	I	Y	D	R	S
U	E	D	W	N	H	E	E	G	R	Y	S
J	D	S	E	B	M	S	D	D	Y	I	T
P	L	A	Y	E	D	J	K	R	I	N	U
L	O	P	W	P	L	A	Y	I	N	G	D
A	H	U	R	R	I	E	S	E	G	P	I
Y	X	V	D	J	F	M	W	D	A	Q	E
S	T	U	D	Y	I	N	G	S	C	Y	S

At Home: Review the Word Study Steps to help your child spell new words.

© Macmillan/McGraw-Hill

Name _____

tries	dries	hurries	studies	plays
tried	dried	hurried	studied	played
trying	drying	hurrying	studying	playing

Pattern Power!

Write the spelling words that have one syllable.

1. _____   3. _____   5. _____

2. _____   4. _____   6. _____

Write the spelling words that have two syllables.

7. _____   10. _____   12. _____

8. _____   11. _____   13. _____

9. _____

Write the spelling words that have three syllables.

14. _____

15. _____

Rhyme Time

Write a spelling word that rhymes with each of these words.

16. fries _____

17. stayed _____

18. praying _____

19. buddies _____

20. flying _____

Name_____

tries	dries	hurries	studies	plays
tried	dried	hurried	studied	played
trying	drying	hurrying	studying	playing

What's the Word?

Complete each sentence with a spelling word.

1. The baseball player _____ to get to first base.

2. After school she dances to music and _____ volleyball with friends.

3. I _____ hard for the test, so I hope I did well.

4. We were all _____ to win the race.

5. The socks that were _____ on the line were blowing in the wind.

6. The children _____ in the backyard after lunch.

7. The teacher was hoping that the students were _____.

8. The cat _____ to climb the tree, but she could not do it.

9. Gretel was dropping crumbs while _____ down the path.

10. The machine _____ the clothes by blowing air on them.

Find the Base Words

Write the base word of each spelling word.

11. tries _____

12. playing _____

13. studies _____

14. hurries _____

15. dried _____

Name _____

There are six spelling mistakes in this paragraph. Circle the misspelled words. Write the words correctly on the lines below.

Our class is studieing dances performed by people around the world. People dance for many reasons: to celebrate good things, to welcome visitors, or just to have fun.

We watched videos of children who were plaing and dancing with their friends. They tryed to jump as high and spin as fast as they could. It looked like fun!

I like to dance, too. I take ballet lessons. I am almost always late for class. Sometimes my tights have not dryed out from being washed. Other times I am hurrieing to finish my homework. My sister studys tap dancing. When I get older, I will learn other kinds of dancing, too.

I know why people all over the world dance. It's fun, and it's good exercise!

1. _____   4. _____

2. _____   5. _____

3. _____   6. _____

Writing Activity

Imagine that you are the coach of a soccer team. Write the speech that you would give your players before the big game. Use at least four spelling words in your description.

_____

_____

_____

_____

Name_____

Look at the words in each set below. One word in each set
is spelled correctly. Look at Sample A. The letter next to the
correctly spelled word in Sample A has been shaded in. Do
Sample B yourself. Shade the letter of the word that is spelled
correctly. When you are sure you know what to do, go on with
the rest of the page.

Sample A:
- Ⓐ cryed
- Ⓑ cried
- Ⓒ cride
- Ⓓ creid

Sample B:
- Ⓔ crys
- Ⓕ chries
- Ⓖ crise
- Ⓗ cries

1.
- Ⓐ trize
- Ⓑ tries
- Ⓒ trys
- Ⓓ treis

2.
- Ⓔ tryed
- Ⓕ tride
- Ⓖ tryd
- Ⓗ tried

3.
- Ⓐ trying
- Ⓑ trian
- Ⓒ treyeing
- Ⓓ triying

4.
- Ⓔ drys
- Ⓕ dryes
- Ⓖ drize
- Ⓗ dries

5.
- Ⓐ dryed
- Ⓑ dreid
- Ⓒ dried
- Ⓓ dride

6.
- Ⓔ driing
- Ⓕ drieing
- Ⓖ drying
- Ⓗ dring

7.
- Ⓐ hurrees
- Ⓑ hurrys
- Ⓒ hurries
- Ⓓ huries

8.
- Ⓔ hurried
- Ⓕ hureed
- Ⓖ hurryed
- Ⓗ huried

9.
- Ⓔ hurriing
- Ⓕ hurrying
- Ⓖ hurryin
- Ⓗ herrying

10.
- Ⓐ studies
- Ⓑ studyes
- Ⓒ studys
- Ⓓ studees

11.
- Ⓔ studyed
- Ⓕ studied
- Ⓖ studdied
- Ⓗ studeed

12.
- Ⓐ studing
- Ⓑ studieing
- Ⓒ studeeing
- Ⓓ studying

13.
- Ⓐ plaze
- Ⓑ plays
- Ⓒ plaise
- Ⓓ plais

14.
- Ⓔ plaide
- Ⓕ playde
- Ⓖ playd
- Ⓗ played

15.
- Ⓐ playing
- Ⓑ plaing
- Ⓒ plaeng
- Ⓓ playeing

Name_____

Fold back the paper along the dotted line. Use the blanks to write each word as it is read aloud. When you finish the test, unfold the paper. Use the list at the right to correct any spelling mistakes.

1. _____
2. _____
3. _____
4. _____
5. _____
6. _____
7. _____
8. _____
9. _____
10. _____
11. _____
12. _____
13. _____
14. _____
15. _____

Review Words 16. _____
17. _____
18. _____

Challenge Words 19. _____
20. _____

1. basket
2. rabbit
3. napkin
4. letter
5. invite
6. bedtime
7. mammal
8. number
9. fellow
10. chapter
11. follow
12. problem
13. chicken
14. butter
15. Sunday
16. tried
17. studies
18. drying
19. splendid
20. complete

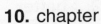

At Home: Help your child practice the words he or she missed to prepare for the Posttest.

The Printer • Book 2/Unit 5 **147**

Name_____

Using the Word Study Steps

1. LOOK at the word.

2. SAY the word aloud.

3. STUDY the letters in the word.

4. WRITE the word.

5. CHECK the word.
   Did you spell the word right?
   If not, go back to step 1.

Find Rhyming Words

Circle the word in each row that rhymes with the word in dark type.

1.	**swallow**	fellow	follow	swell
2.	**better**	letter	ladder	batter
3.	**flutter**	batter	flatter	butter
4.	**yellow**	fellow	yelled	follow
5.	**polite**	police	invite	polish
6.	**camel**	mammal	maple	manage
7.	**habit**	rapid	rather	rabbit
8.	**Monday**	birthday	today	Sunday

X the Word

Put an X on the word in each row that does not fit the pattern.

9.	napkin	basket	rabbit	tried
10.	Sunday	drying	number	butter
11.	studies	follow	problem	stopping
12.	chicken	bedtime	click	jumping

 At Home: Review the Word Study Steps to help your child spell new words.

Name _____

basket	letter	mammal	chapter	chicken
rabbit	invite	number	follow	butter
napkin	bedtime	fellow	problem	Sunday

Pattern Power!

Write the spelling words with these spelling patterns.

_ll_

1. _____

2. _____

_mm_

3. _____

_tt_

4. _____

5. _____

_bb_

6. _____

Order Please!

Write each group of spelling words in alphabetical order.

basket, chapter, chicken, bedtime

7. _____

8. _____

9. _____

10. _____

invite, napkin, Sunday, number, problem

11. _____

12. _____

13. _____

14. _____

15. _____

Name_____

basket	letter	mammal	chapter	chicken
rabbit	invite	number	follow	butter
napkin	bedtime	fellow	problem	Sunday

What's the Word?

Complete each sentence with a spelling word.

1. The nurse brought the sick man soup and bread with _____.

2. The police officers' awards dinner will be on _____ night.

3. The next _____ of the book is about firefighters.

4. My mom rescued my pet _____ from a neighborhood dog.

5. Officer Dan is a wonderful father and a good _____.

6. When you have a _____, ask your parents for help.

7. Will you _____ your friend to the firehouse fundraiser?

8. _____ the police officer to safety.

9. They delivered a _____ of fruit to the firefighters who saved them.

10. I wrote a _____ to her because she is my hero.

It Takes Three

Write a spelling word that goes with the other two words.

11. reptile, bird, _____

12. fork, placemat, _____

13. letter, symbol, _____

14. pig, cow, _____

15. lunchtime, daytime, _____

Name_____

**There are six spelling mistakes in this story. Circle the
misspelled words. Write the words correctly on the lines below.**

The Great Rescue

One Sonday afternoon my family went on a picnic. My mother packed
our lunch in a big bascett. We found a shady spot under a tree and ate fried
chikin and bread with bauter.

After lunch my sister and I took a canoe out on the lake. We were drifting
along when suddenly my sister screamed. I looked over to see what the
problime was. There was a hole in the bottom of the boat. The canoe was
filling up with water.

Luckily our parents saw us and ran for help. Another person said he
would help. This fine fello swam out to the canoe. He held out a rope for us
to grab onto, and then he dragged us to shore. He was our hero!

It was an exciting end to an almost peaceful picnic.

1. _____   4. _____

2. _____   5. _____

3. _____   6. _____

Writing Activity

**Imagine that you are on a picnic and something unexpected
happens. Write a short story about it, using at least four
spelling words.**

_____

_____

_____

_____

_____

Look at the words in each set below. One word in each set is spelled correctly. Look at Sample A. The letter next to the correctly spelled word in Sample A has been shaded in. Do Sample B yourself. Shade the letter of the word that is spelled correctly. When you are sure you know what to do, go on with the rest of the page.

Sample A:

Ⓐ ento
Ⓑ intoo
Ⓒ into
Ⓓ intue

Sample B:

Ⓔ daddy
Ⓕ daddie
Ⓖ dady
Ⓗ dadie

1. Ⓐ basket
   Ⓑ bakset
   Ⓒ baseket
   Ⓓ basskit

2. Ⓔ rabbit
   Ⓕ rabit
   Ⓖ rabbet
   Ⓗ rabet

3. Ⓐ nappkin
   Ⓑ napkin
   Ⓒ napekin
   Ⓓ napkine

4. Ⓔ leter
   Ⓕ lettar
   Ⓖ letar
   Ⓗ letter

5. Ⓐ invite
   Ⓑ envite
   Ⓒ invit
   Ⓓ inevite

6. Ⓔ bedtim
   Ⓕ bedtime
   Ⓖ beddtime
   Ⓗ beadtime

7. Ⓐ mamal
   Ⓑ mammel
   Ⓒ mammal
   Ⓓ mamul

8. Ⓔ nummer
   Ⓕ numer
   Ⓖ numbur
   Ⓗ number

9. Ⓔ felow
   Ⓕ fellow
   Ⓖ felou
   Ⓗ felloou

10. Ⓐ chapper
    Ⓑ chapter
    Ⓒ chatper
    Ⓓ chappter

11. Ⓔ folow
    Ⓕ follo
    Ⓖ follow
    Ⓗ falow

12. Ⓐ probem
    Ⓑ problum
    Ⓒ problem
    Ⓓ prolbem

13. Ⓐ chicken
    Ⓑ chiken
    Ⓒ chiccen
    Ⓓ chickn

14. Ⓔ buter
    Ⓕ butter
    Ⓖ buttar
    Ⓗ butor

15. Ⓐ Senday
    Ⓑ Sundai
    Ⓒ Sudnay
    Ⓓ Sunday

© Macmillan/McGraw-Hill

Name _____

Fold back the paper along the dotted line. Use the blanks to write each word as it is read aloud. When you finish the test, unfold the paper. Use the list at the right to correct any spelling mistakes.

1. _____
2. _____
3. _____
4. _____
5. _____
6. _____
7. _____
8. _____
9. _____
10. _____
11. _____
12. _____
13. _____
14. _____
15. _____

Review Words 16. _____
17. _____
18. _____

Challenge Words 19. _____
20. _____

1. pilot
2. diner
3. tiger
4. favor
5. lemon
6. planet
7. model
8. shady
9. robot
10. tiny
11. label
12. cozy
13. silent
14. spider
15. frozen
16. follow
17. basket
18. Sunday
19. melon
20. stomach

At Home: Help your child practice the words he or she missed to prepare for the Posttest.

Name

Using the Word Study Steps

1. LOOK at the word.
2. SAY the word aloud.
3. STUDY the letters in the word.

4. WRITE the word.
5. CHECK the word.
   Did you spell the word right?
   If not, go back to step 1.

Find and Circle

Where are the spelling words?

S	P	I	D	E	R	C	O	Z	Y
Z	G	W	E	M	O	D	E	L	G
X	Q	P	O	B	B	H	J	U	T
F	N	F	C	V	O	V	F	J	I
P	L	A	N	E	T	N	T	R	G
I	A	V	T	L	Y	K	I	L	E
L	B	O	F	R	D	I	N	E	R
O	E	R	S	H	A	D	Y	M	E
T	L	S	I	L	E	N	T	O	F
U	O	P	F	R	O	Z	E	N	X

At Home: Review the Word Study Steps to help your child spell new words.

Name _____

pilot	favor	model	tiny	silent
diner	lemon	shady	label	spider
tiger	planet	robot	cozy	frozen

Rhyme Time

Write the spelling word that rhymes with each word below.

1. chosen _____

2. table _____

3. flavor _____

4. lady _____

5. nosy _____

6. rider _____

Syllable Patterns

How a word is divided into syllables may depend on whether the vowel in the first part of the word is long or short. If the first syllable has a short vowel sound, it is usually divided after the consonant. If the first syllable has a long vowel sound, it is usually divided after the vowel. Divide each spelling word into syllables.

7. pilot _____

8. planet _____

9. robot _____

10. model _____

11. diner _____

12. silent _____

13. tiger _____

14. lemon _____

15. tiny _____

Name_____

pilot	favor	model	tiny	silent
diner	lemon	shady	label	spider
tiger	planet	robot	cozy	frozen

What's the Word?

Complete each sentence with a spelling word.

1. The black _____ was spinning a new web.

2. The large, _____ oak tree is home to many animals.

3. The Bengal _____, which is a kind of large cat, lives in India.

4. Birds build nests on every part of the _____ Earth.

5. Many fish live beneath the surface of a _____ pond.

6. The black bear slept in his _____, warm den all winter.

7. It takes many _____ ants to build an anthill.

8. Bugs were living in the _____ tree and eating the sour fruit.

9. The owl was as _____ as a mouse as she landed in her nest.

10. I made a clay _____ of a beehive for my science project.

Define It!

Write the spelling word that has the same meaning as each word or phrase below.

11. A tag _____

12. A machine that looks like a person _____

13. A small restaurant _____

14. Special help given to a friend _____

15. A person who steers a plane _____

There are six spelling mistakes in this report. Circle the misspelled words. Write the words correctly on the lines below.

King of the Jungle

Some people think that the lion is the king of the jungle, but lions do not even live in forests. The real king of the jungle is the tigger!

I did a report on these large cats. Here are a few things I learned. These animals live in Asia, not in Africa, as many people think. They are tiene when they are born, only 2 or 3 pounds, but they grow fast. The biggest one on the plannett weighs more than 1,000 pounds! Because they live alone, they can be siellent as they walk through the shadey forest looking for food.

The next time someone calls a lion the king of the jungle, do me a faiver, and tell them the truth!

1. _____   4. _____

2. _____   5. _____

3. _____   6. _____

Writing Activity

Imagine that you are an insect living in your backyard. Write a paragraph describing something about your life in this backyard home. Use at least four spelling words in your description.

_____

_____

_____

_____

_____

Name_____

Look at the words in each set below. One word in each set is spelled correctly. Look at Sample A. The letter next to the correctly spelled word in Sample A has been shaded in. Do Sample B yourself. Shade the letter of the word that is spelled correctly. When you are sure you know what to do, go on with the rest of the page.

Sample A:

Ⓐ wagon
Ⓑ wagen
Ⓒ waggon
Ⓓ waggen

Sample B:

Ⓔ pallace
Ⓕ pallase
Ⓖ palase
Ⓗ palace

1. Ⓐ pilote
   Ⓑ pilot
   Ⓒ pillot
   Ⓓ pielot

2. Ⓔ dinar
   Ⓕ dinnar
   Ⓖ dyner
   Ⓗ diner

3. Ⓐ tiger
   Ⓑ tigger
   Ⓒ tyger
   Ⓓ tiggur

4. Ⓔ favur
   Ⓕ favvor
   Ⓖ favor
   Ⓗ favore

5. Ⓐ lemone
   Ⓑ lemon
   Ⓒ lemmon
   Ⓓ lemmin

6. Ⓔ planet
   Ⓕ plannet
   Ⓖ planit
   Ⓗ plannit

7. Ⓐ modell
   Ⓑ model
   Ⓒ moddel
   Ⓓ modul

8. Ⓔ shadey
   Ⓕ shadie
   Ⓖ shady
   Ⓗ shadee

9. Ⓔ robbut
   Ⓕ robat
   Ⓖ robot
   Ⓗ robbat

10. Ⓐ tinee
    Ⓑ tinnie
    Ⓒ tiney
    Ⓓ tiny

11. Ⓔ labbel
    Ⓕ labil
    Ⓖ label
    Ⓗ labul

12. Ⓐ cozzy
    Ⓑ cozie
    Ⓒ cosie
    Ⓓ cozy

13. Ⓐ silente
    Ⓑ silent
    Ⓒ sillent
    Ⓓ silant

14. Ⓔ spidur
    Ⓕ spiddar
    Ⓖ spider
    Ⓗ spidder

15. Ⓐ frozen
    Ⓑ frowzen
    Ⓒ frausen
    Ⓓ frauzen

© Macmillan/McGraw-Hill

Name_____

Read each sentence. If an underlined word is spelled wrong, fill in the circle that goes with that word. If no word is spelled wrong, fill in the circle below NONE. Read Sample A, and do Sample B.

NONE

A. I want to <u>see</u> the <u>whales</u> very <u>mutch</u>.
        A   B   C

A. Ⓐ Ⓑ Ⓒ Ⓓ

NONE

B. The <u>truth</u> is that the <u>nite</u> should be on the <u>throne</u>.
      E   F   G

B. Ⓔ Ⓕ Ⓖ Ⓗ

NONE

1. He <u>studies</u> the math <u>problime</u> in his <u>notebook</u>.
      A   B   C

1. Ⓐ Ⓑ Ⓒ Ⓓ

NONE

2. The <u>tiney</u> bear cub <u>tried</u> to climb the <u>frozen</u> hill.
    E   F   G

2. Ⓔ Ⓕ Ⓖ Ⓗ

NONE

3. <u>Follow</u> the cat to its <u>cozy</u> hiding place in the <u>shadey</u> tree.
    A   B   C

3. Ⓐ Ⓑ Ⓒ Ⓓ

NONE

4. <u>Hurrieing</u> down the <u>frozen</u> path, he <u>dropped</u> his scarf.
    E   F   G

4. Ⓔ Ⓕ Ⓖ Ⓗ

NONE

5. I read about the <u>airplane</u> in the <u>newspaper</u> last <u>Sonday</u>.
         A   B   C

5. Ⓐ Ⓑ Ⓒ Ⓓ

NONE

6. She <u>played</u> the piano while the <u>tiny</u> kids were <u>danceing</u>.
      E   F   G

6. Ⓔ Ⓕ Ⓖ Ⓗ

NONE

7. <u>Sumone</u> was <u>drying</u> his wet socks by the <u>cozy</u> fireplace.
    A   B   C

7. Ⓐ Ⓑ Ⓒ Ⓓ

NONE

8. She <u>hoped</u> to sit beneath the <u>shady</u> tree and stay <u>silent</u>.
      E   F   G

8. Ⓔ Ⓕ Ⓖ Ⓗ

NONE

9. We were <u>hurrying</u> to catch our <u>areplan</u> when we
          A   B

heard there was a <u>problem</u>.
             C

9. Ⓐ Ⓑ Ⓒ Ⓓ

NONE

10. <u>Someone</u> put the <u>letere</u> in my <u>notebook</u>.
      E   F   G

10. Ⓔ Ⓕ Ⓖ Ⓗ

NONE

11. I was <u>drying</u> my eyes as silent tears <u>droped</u> from my chin.
      A   B   C

11. Ⓐ Ⓑ Ⓒ Ⓓ

Name_____

NONE
12. The <u>newspaper</u> <u>names</u> who is having a <u>burtheday</u>.   12. Ⓔ Ⓕ Ⓖ Ⓗ
               E   F   G

NONE
13. I was <u>wraping</u> the <u>birthday</u> gift in <u>newspaper</u>.   13. Ⓐ Ⓑ Ⓒ Ⓓ
            A   B   C

NONE
14. He is <u>sillint</u> while he <u>studies</u> the <u>letter</u>.   14. Ⓔ Ⓕ Ⓖ Ⓗ
          E   F   G

NONE
15. The <u>tiny</u> <u>airplane</u> landed on the <u>frozzen</u> ground.   15. Ⓐ Ⓑ Ⓒ Ⓓ
         A   B   C

NONE
16. We <u>plaied</u> and <u>hoped</u> the rain would stop, while our   16. Ⓔ Ⓕ Ⓖ Ⓗ
        E   F

boots were <u>drying</u>.
             G

NONE
17. We were <u>hurrying</u> as we <u>tried</u> to <u>folow</u> the guide.   17. Ⓐ Ⓑ Ⓒ Ⓓ
             A   B   C

NONE
18. She <u>studys</u> steps in the <u>notebook</u>, then begins <u>dancing</u>.   18. Ⓔ Ⓕ Ⓖ Ⓗ
       E   F   G

NONE
19. The teacher <u>namms</u> <u>someone</u> to solve the <u>problem</u>.   19. Ⓐ Ⓑ Ⓒ Ⓓ
               A   B   C

NONE
20. We <u>tryed</u> to <u>follow</u> the directions in the <u>letter</u>.   20. Ⓔ Ⓕ Ⓖ Ⓗ
        E   F   G

NONE
21. <u>Dancing</u> under the <u>shady</u> tree, I <u>hopd</u> it wouldn't rain.   21. Ⓐ Ⓑ Ⓒ Ⓓ
    A   B   C

NONE
22. He <u>played</u> while she was <u>wrapping</u> the gift <u>bascit</u>.   22. Ⓔ Ⓕ Ⓖ Ⓗ
       E   F   G

NONE
23. We gave her a <u>basket</u> for her <u>birthday</u> on <u>Sunday</u>.   23. Ⓐ Ⓑ Ⓒ Ⓓ
             A   B   C

NONE
24. While <u>wrapping</u> presents in the <u>cosey</u> room, she   24. Ⓔ Ⓕ Ⓖ Ⓗ
         E   F

<u>names</u> winter her favorite season.
 G

NONE
25. We <u>dropped</u> off a <u>basket</u> for the brunch on <u>Sunday</u>.   25. Ⓐ Ⓑ Ⓒ Ⓓ
    A   B   C

Fold back the paper along the dotted line. Use the blanks to write each word as it is read aloud. When you finish the test, unfold the paper. Use the list at the right to correct any spelling mistakes.

1. _____
2. _____
3. _____
4. _____
5. _____
6. _____
7. _____
8. _____
9. _____
10. _____
11. _____
12. _____
13. _____
14. _____
15. _____

Review Words 16. _____

17. _____

18. _____

Challenge Words 19. _____

20. _____

1. able
2. purple
3. riddle
4. handle
5. towel
6. eagle
7. puzzle
8. castle
9. little
10. nickel
11. camel
12. pickle
13. travel
14. tunnel
15. squirrel
16. spider
17. tiny
18. planet
19. motel
20. couple

At Home: Help your child practice the words he or she missed to prepare for the Posttest.

Name _____

Using the Word Study Steps

1. LOOK at the word.

2. SAY the word aloud.

3. STUDY the letters in the word.

4. WRITE the word.

5. CHECK the word.
Did you spell the word right?
If not, go back to step 1.

Rhyme Time

Circle the word in each row that rhymes with the word in dark type.

1.	**tickle**	ticket	pickle	picnic
2.	**table**	able	bubble	trouble
3.	**middle**	rattle	puddle	riddle
4.	**owl**	towel	bowl	town
5.	**pickle**	picking	nickel	nibble
6.	**candle**	handy	handsome	handle

X the Word

Put an X on the word in each row that does not fit the pattern.

7.	little	puzzle	barrel	riddle
8.	pickle	panel	purple	puzzle
9.	travel	tunnel	towel	handle
10.	castle	camel	cattle	couple
11.	awful	able	eagle	table
12.	handle	stable	sparkle	squirrel

© Macmillan/McGraw-Hill

At Home: Review the Word Study Steps to help your child spell new words.

able	handle	puzzle	nickel	travel
purple	towel	castle	camel	tunnel
riddle	eagle	little	pickle	squirrel

Pattern Power!

Write the spelling words with these spelling patterns for the final /əl/ sound.

-el

1. _____   4. _____

2. _____   5. _____

3. _____   6. _____

-le

7. _____   12. _____

8. _____   13. _____

9. _____   14. _____

10. _____   15. _____

11. _____

Syllable Patterns

How a word is divided into syllables may depend on whether the vowel in the first part of the word is long or short. Divide the following spelling words into syllables.

16. able _____   19. nickel _____

17. camel _____   20. handle _____

18. eagle _____

Name_____

able	handle	puzzle	nickel	travel
purple	towel	castle	camel	tunnel
riddle	eagle	little	pickle	squirrel

What's the Word?

Complete each sentence with a spelling word.

1. Our class was _____ to help out at the soup kitchen.

2. Save an extra dime or _____ each week to give to help others.

3. The doctors _____ to other countries to help the sick.

4. The house is not a _____, but it is a nice home.

5. If we each give a _____ bit of ourselves, our city can be stronger.

6. We drive through a _____ on our way to the homeless shelter.

It Takes Three

Read the heading for each group of words. Then add the spelling word that belongs in that group.

Birds

7. owl, hawk, _____

Colors

8. green, orange, _____

Things Used for Washing Hands

9. soap, water, _____

Desert Animals

10. lizard, coyote, _____

© Macmillan/McGraw-Hill

Name_____

There are six spelling mistakes in this paragraph. Circle the misspelled words. Write the words correctly on the lines below.

My parents always taught me that it is important to give to those who have less than we do. It does not have to mean giving money. If you are abbel, it is also great to give your time and talents.

Each summer, we travle to a nearby city and work with a group of people building a house for another family. Even when I was litel, I could help by passing out lunch to the workers. Everyone got a sandwich and a pikkel. Then I would spend the afternoon playing with toys or a puzle with other kids.

Last year, I was old enough to handel a bigger job. I would bring the right tools to my mom and dad.

It feels good to help others. It can be a lot of fun too!

1. _____   4. _____

2. _____   5. _____

3. _____   6. _____

Writing Activity

Do you have a hero? People who help others are community heroes. Write four questions you would like to ask your hero. Use at least three spelling words in your questions.

_____

_____

_____

_____

_____

_____

Name _____

Look at the words in each set below. One word in each set is spelled correctly. Look at Sample A. The letter next to the correctly spelled word in Sample A has been shaded in. Do Sample B yourself. Shade the letter of the word that is spelled correctly. When you are sure you know what to do, go on with the rest of the page.

Sample A:

- Ⓐ funnel
- Ⓑ funnul
- Ⓒ funnle
- Ⓓ funnal

Sample B:

- Ⓔ stabal
- Ⓕ stable
- Ⓖ stabel
- Ⓗ stabble

1. Ⓐ abel
   Ⓑ abal
   Ⓒ abul
   Ⓓ able

2. Ⓔ purple
   Ⓕ purpal
   Ⓖ purpel
   Ⓗ purpple

3. Ⓐ riddel
   Ⓑ riddle
   Ⓒ riddal
   Ⓓ riddull

4. Ⓔ handle
   Ⓕ handel
   Ⓖ handal
   Ⓗ handell

5. Ⓐ towle
   Ⓑ towwell
   Ⓒ towell
   Ⓓ towel

6. Ⓔ eagel
   Ⓕ eagal
   Ⓖ eagell
   Ⓗ eagle

7. Ⓐ puzzel
   Ⓑ puzzle
   Ⓒ puzzal
   Ⓓ puzall

8. Ⓔ castel
   Ⓕ cassel
   Ⓖ castal
   Ⓗ castle

9. Ⓐ littel
   Ⓑ little
   Ⓒ litle
   Ⓓ littal

10. Ⓔ nickle
    Ⓕ nickal
    Ⓖ nickel
    Ⓗ nickell

11. Ⓐ cammel
    Ⓑ camel
    Ⓒ cammle
    Ⓓ camle

12. Ⓔ pickel
    Ⓕ pickal
    Ⓖ pickle
    Ⓗ pickul

13. Ⓐ travle
    Ⓑ traval
    Ⓒ travel
    Ⓓ travvell

14. Ⓔ tunnel
    Ⓕ tunnle
    Ⓖ tunnal
    Ⓗ tunell

15. Ⓐ squirl
    Ⓑ squirrel
    Ⓒ squirrul
    Ⓓ squirel

Name

Fold back the paper along the dotted line. Use the blanks to write each word as it is read aloud. When you finish the test, unfold the paper. Use the list at the right to correct any spelling mistakes.

1. _____
2. _____
3. _____
4. _____
5. _____
6. _____
7. _____
8. _____
9. _____
10. _____
11. _____
12. _____
13. _____
14. _____
15. _____

Review Words 16. _____

17. _____

18. _____

Challenge Words 19. _____

20. _____

1. untied
2. repay
3. disagree
4. preheat
5. unafraid
6. return
7. preschool
8. dislike
9. disappear
10. resell
11. precook
12. prepay
13. unbeaten
14. reprint
15. unwrap
16. nickel
17. handle
18. pickle
19. unlucky
20. recover

© Macmillan/McGraw-Hill

At Home: Help your child practice the words he or she missed to prepare for the Posttest.

Wilbur's Boast • Book 2/Unit 6   167

Using the Word Study Steps

1. LOOK at the word.

2. SAY the word aloud.

3. STUDY the letters in the word.

4. WRITE the word.

5. CHECK the word.
   Did you spell the word right?
   If not, go back to step 1.

Find and Circle

Find and circle the hidden spelling words.

D	I	S	A	P	P	E	A	R	U
I	U	X	U	R	R	R	X	E	N
S	N	Z	N	E	E	E	D	P	A
A	T	K	W	P	S	T	I	R	F
G	I	Q	R	A	C	U	S	I	R
R	E	P	A	Y	H	R	L	N	A
E	D	Q	P	Z	O	N	I	T	I
E	P	R	E	C	O	O	K	Q	D
R	E	S	E	L	L	K	E	X	Z
U	N	B	E	A	T	E	N	Z	K
K	X	P	R	E	H	E	A	T	Z

 At Home: Review the Word Study Steps to help your child spell new words.

© Macmillan/McGraw-Hill

Name_____

untied	preheat	preschool	resell	unbeaten
repay	unafraid	dislike	precook	reprint
disagree	return	disappear	prepay	unwrap

Write the spelling words that begin with each of these prefixes.

re-

1. _____

2. _____

3. _____

4. _____

dis-

5. _____

6. _____

7. _____

un-

8. _____

9. _____

10. _____

11. _____

pre-

12. _____

13. _____

14. _____

15. _____

Name_____

untied	preheat	preschool	resell	unbeaten
repay	unafraid	dislike	precook	reprint
disagree	return	disappear	prepay	unwrap

What's the Word?

Complete each sentence with a spelling word.

1. I do not know anyone who would _____ my puppy.

2. She _____ his leash and took him for a walk.

3. Can we _____ the snake to the pet store?

4. My mother will _____ the oven before baking the cake.

5. The ducks _____ under the water.

6. I have to _____ with you—I think pets are terrific!

7. She brought her pet hamster with her to _____.

8. We can never _____ you for saving our cat.

9. My turtle is _____ in our neighborhood turtle races.

10. That rabbit knows how to _____ a candy bar!

Find the Base

Write the base word for the spelling words below.

11. unafraid _____

12. prepay _____

13. resell _____

14. reprint _____

15. precook _____

Name_____

There are six spelling mistakes in this paragraph. Circle the misspelled words. Write the words correctly on the lines below.

My sister Ariel is in preskoul. She has a cat named Sally. Ariel and Sally both disslike dogs. They run screaming if they see a dog, no matter how gentle or sweet the dog is.

One day, Ariel and I were in the yard playing when we heard a dog barking near our house. I was unafrade, so I went to see why he was barking. When I got closer, I saw Sally tangled in the dog's leash. She was scratching at him, but he was just sitting there calling for help. When I untyed the cat, the dog ran off.

Ariel would like to reepay the dog for his kindness, but we have never seen him rettern.

1. _____   4. _____

2. _____   5. _____

3. _____   6. _____

Writing Activity

Do you have a pet? Is there an animal you would like to have for a pet? Write a paragraph describing your pet or the pet you would like to have. Use at least four spelling words in your description.

_____

_____

_____

_____

_____

Name_____

Look at the words in each set below. One word in each set is spelled correctly. Look at Sample A. The letter next to the correctly spelled word in Sample A has been shaded in. Do Sample B yourself. Shade the letter of the word that is spelled correctly. When you are sure you know what to do, go on with the rest of the page.

Sample A:

Ⓐ rerrun
Ⓑ reerun
Ⓒ rerun
Ⓓ rerunn

Sample B:

Ⓔ unwind
Ⓕ unwinde
Ⓖ unwynd
Ⓗ unwynde

1. Ⓐ untied
   Ⓑ entied
   Ⓒ unteid
   Ⓓ uhntied

2. Ⓔ reepay
   Ⓕ repaye
   Ⓖ repay
   Ⓗ repai

3. Ⓐ dissagree
   Ⓑ disagree
   Ⓒ disagre
   Ⓓ dissagrea

4. Ⓔ preheat
   Ⓕ preet
   Ⓖ preeat
   Ⓗ perheat

5. Ⓐ unfraid
   Ⓑ unnafraid
   Ⓒ unafrade
   Ⓓ unafraid

6. Ⓔ ruhturn
   Ⓕ return
   Ⓖ ruturn
   Ⓗ retern

7. Ⓐ preschol
   Ⓑ preschool
   Ⓒ preeschool
   Ⓓ preskool

8. Ⓔ disslike
   Ⓕ dislaik
   Ⓖ disslake
   Ⓗ dislike

9. Ⓐ disappear
   Ⓑ disapear
   Ⓒ dissapear
   Ⓓ disappeer

10. Ⓔ resell
    Ⓕ resel
    Ⓖ reesel
    Ⓗ reesell

11. Ⓐ percook
    Ⓑ prekook
    Ⓒ preecok
    Ⓓ precook

12. Ⓔ perpay
    Ⓕ prepai
    Ⓖ preepay
    Ⓗ prepay

13. Ⓐ unbeeten
    Ⓑ unbeatan
    Ⓒ unbeaten
    Ⓓ unbieten

14. Ⓔ repprint
    Ⓕ repint
    Ⓖ reprint
    Ⓗ reeprint

15. Ⓐ unrapp
    Ⓑ unwrap
    Ⓒ unwrapp
    Ⓓ unrap

Name _____

Fold back the paper along the dotted line. Use the blanks to write each word as it is read aloud. When you finish the test, unfold the paper. Use the list at the right to correct any spelling mistakes.

1. _____   **1.** sister

2. _____   **2.** sailor

3. _____   **3.** dollar

4. _____   **4.** toaster

5. _____   **5.** winter

6. _____   **6.** doctor

7. _____   **7.** later

8. _____   **8.** dancer

9. _____   **9.** mayor

10. _____   **10.** writer

11. _____   **11.** silver

12. _____   **12.** cellar

13. _____   **13.** trailer

14. _____   **14.** December

15. _____   **15.** author

Review Words   16. _____   **16.** resell

17. _____   **17.** prepay

18. _____   **18.** unwrap

Challenge Words   19. _____   **19.** circular

20. _____   **20.** editor

© Macmillan/McGraw-Hill

At Home: Help your child practice the words he or she missed to prepare for the Posttest.

An American Hero Flies Again   **173**

Book 2/Unit 6

Name_____

Using the Word Study Steps

1. LOOK at the word.

2. SAY the word aloud.

3. STUDY the letters in the word.

4. WRITE the word.

5. CHECK the word.
   Did you spell the word right?
   If not, go back to step 1.

Rhyme Time

Circle the word in each row that rhymes with the word in dark type.

1.	**greater**	later	letter	lighter
2.	**trailer**	silver	sailing	sailor
3.	**fighter**	writer	water	fatter
4.	**collar**	color	dollar	deliver
5.	**answer**	drawer	dancer	dinner
6.	**mister**	sister	master	clowns

X the Word

Put an X on the word in each row that does not fit the pattern.

7.	sister	silver	summer	sailor
8.	December	dancer	doctor	danger
9.	tractor	trailer	toaster	temper
10.	cellar	dollar	duller	collar
11.	winter	visitor	writer	later
12.	matter	author	mayor	anchor

At Home: Review the Word Study Steps to help your child spell new words.

Name_____

sister	toaster	later	writer	trailer
sailor	winter	dancer	silver	December
dollar	doctor	mayor	cellar	author

Pattern Power!

Write the spelling words with these spelling patterns for the final /ər/ sound.

-er

1. _____
2. _____
3. _____
4. _____
5. _____
6. _____
7. _____
8. _____
9. _____

-or

10. _____
11. _____
12. _____
13. _____

-ar

14. _____
15. _____

Syllable Patterns

How a word is divided into syllables may depend on whether the vowel in the first part of the word is long or short. Divide the following spelling words into syllables.

16. sailor _____

17. winter _____

18. toaster _____

19. author _____

20. trailer _____

Name_____

sister	toaster	later	writer	trailer
sailor	winter	dancer	silver	December
dollar	doctor	mayor	cellar	author

Finish the Sentence

Complete each sentence using a spelling word.

1. This year, my oldest _____ will be able to vote.

2. Thomas Jefferson was one _____ of the Declaration of Independence.

3. My parents went to a rally for the _____.

4. Elections take place _____ in the year.

5. He started his campaign in the month of _____.

6. In the _____, it is important to donate blankets and clothing to shelters.

7. George Washington's picture is on the _____ bill.

8. The candidate travels in a _____ during the campaign.

Analogies

An analogy is a statement that compares sets of words that are alike in some way: *Night* is to *day* as *black* is to *white*. This analogy points out that *night* and *day* are opposite in the same way that *black* and *white* are opposite.

Use the spelling words to complete the analogies below.

1. *Jazz* is to *musician* as *ballet* is to _____ .

2. *Coffee* is to *pot* as *bread* is to _____.

3. *Yellow* is to *gold* as *gray* is to _____.

5. *Brush* is to *painter* as *pen* is to _____.

There are six spelling mistakes in this speech. Circle the misspelled words. Write the words correctly on the lines below.

   Hello, my fellow citizens. Thank you so much for coming out on this cold winnter day. Seeing you all here is very inspiring. Running for mayur has been a great experience. I have met so many people in this community. Yesterday, I spoke to a docter who is concerned about our hospitals. Latir, I spoke to a sailer who is afraid our ports are too crowded. I even spoke to a danser who says there is not enough appreciation for the arts. With your help we will win this election. Then, I can help everyone with these issues.

   I'll see you on election day!

1. _____ 4. _____

2. _____ 5. _____

3. _____ 6. _____

Writing Activity

Pretend you are running for class president. Write a speech you would give to your class. Use at least three spelling words in your paragraph.

_____

_____

_____

_____

_____

_____

Look at the words in each set below. One word in each set is spelled correctly. Look at Sample A. The letter next to the correctly spelled word in Sample A has been shaded in. Do Sample B yourself. Shade the letter of the word that is spelled correctly. When you are sure you know what to do, go on with the rest of the page.

Sample A:

Ⓐ rivare
Ⓑ rivur
Ⓒ riverr
Ⓓ river

Sample B:

Ⓔ swimmer
Ⓕ swimer
Ⓖ swimmur
Ⓗ swimmir

1. Ⓐ sistur
   Ⓑ sisster
   Ⓒ sister
   Ⓓ sistar

6. Ⓔ doctor
   Ⓕ dokter
   Ⓖ docter
   Ⓗ doctar

11. Ⓐ silvur
    Ⓑ silvher
    Ⓒ silver
    Ⓓ silvor

2. Ⓔ sailor
   Ⓕ sailir
   Ⓖ sailer
   Ⓗ sailar

7. Ⓐ lattar
   Ⓑ latar
   Ⓒ lator
   Ⓓ later

12. Ⓔ sellar
    Ⓕ celler
    Ⓖ cellar
    Ⓗ celar

3. Ⓐ doller
   Ⓑ dollar
   Ⓒ dollor
   Ⓓ dollur

8. Ⓔ dancer
   Ⓕ danser
   Ⓖ dancur
   Ⓗ dancir

13. Ⓐ traylor
    Ⓑ trailor
    Ⓒ traler
    Ⓓ trailer

4. Ⓔ toster
   Ⓕ tosteer
   Ⓖ taoster
   Ⓗ toaster

9. Ⓐ maior
   Ⓑ mayor
   Ⓒ mayore
   Ⓓ mayer

14. Ⓔ Decemeber
    Ⓕ Decembur
    Ⓖ December
    Ⓗ Decembir

5. Ⓐ wintir
   Ⓑ winter
   Ⓒ whinter
   Ⓓ wenter

10. Ⓔ riter
    Ⓕ writer
    Ⓖ writor
    Ⓗ writtor

15. Ⓐ auther
    Ⓑ authur
    Ⓒ author
    Ⓓ arthor

© Macmillan/McGraw-Hill

Name _____

Fold back the paper along the dotted line. Use the blanks to write each word as it is read aloud. When you finish the test, unfold the paper. Use the list at the right to correct any spelling mistakes.

1. _____
2. _____
3. _____
4. _____
5. _____
6. _____
7. _____
8. _____
9. _____
10. _____
11. _____
12. _____
13. _____
14. _____
15. _____

Review Words 16. _____
17. _____
18. _____

Challenge Words 19. _____
20. _____

1. careful
2. cheerful
3. helpful
4. colorful
5. harmful
6. peaceful
7. pitiful
8. painless
9. priceless
10. helpless
11. sleepless
12. rainless
13. helplessly
14. carefully
15. peacefully
16. doctor
17. dollar
18. December
19. wonderful
20. cloudless

© Macmillan/McGraw-Hill

At Home: Help your child practice the words he or she missed to prepare for the Posttest.

Mother to Tigers • Book 2/Unit 6   179

Name_____

Using the Word Study Steps

1. LOOK at the word.

2. SAY the word aloud.

3. STUDY the letters in the word.

4. WRITE the word.

5. CHECK the word.
   Did you spell the word right?
   If not, go back to step 1.

X the Word

Put an X on the word that does not fit the pattern in each row.

1. careful	pitiful	barrel	bashful
2. harmful	panel	mindful	peaceful
3. anxiously	carefully	lovely	towel
4. peacefully	hopeful	awfully	beautifully
5. lost	helpless	speechless	sleepless
6. painless	priceless	rainless	peaceful
7. helplessly	money	happily	luckily
8. bubble	cheerful	helpful	peaceful
9. pitiful	colorful	careful	busily
10. rainless	priceless	pitiful	helpless

© Macmillan/McGraw-Hill

At Home: Review the Word Study Steps to help your child spell new words.

Name_____

careful	colorful	pitiful	helpless	helplessly
cheerful	harmful	painless	sleepless	carefully
helpful	peaceful	priceless	rainless	peacefully

Pattern Power!

Write the spelling words that end with each of these suffixes.

-ful

1. _____
2. _____
3. _____
4. _____
5. _____
6. _____
7. _____

-ly

8. _____
9. _____
10. _____

-less

11. _____
12. _____
13. _____
14. _____
15. _____

Name_____

careful	colorful	pitiful	helpless	helplessly
cheerful	harmful	painless	sleepless	carefully
helpful	peaceful	priceless	rainless	peacefully

For each spelling word below write the base word. The first one is done for you.

1. peaceful   __peace__

2. helplessly   _____

3. cheerful   _____

4. helpful   _____

5. harmful   _____

6. helpless   _____

7. colorful   _____

8. painless   _____

9. carefully   _____

10. rainless   _____

11. careful   _____

12. sleepless   _____

13. pitiful   _____

14. priceless   _____

15. peacefully   _____

Which has a base word that changes _y_ to _i_ when adding a suffix?

16. _____

Name_____

There are six spelling mistakes in this paragraph. Circle the misspelled words. Write the words correctly on the lines below.

From behind the tree, the doctor observed the lion as he slept peacfully by the swamp. The doctor knew that even though the lion looked pieceful, he needed to get some very important medicine. To give the lion the medicine, the doctor would have to be very, very carefull. Lions do not like to be surprised. The shot would be paneless, though. Without it, the lion would be helples to fight the disease that was spreading in the jungle. The doctor knew he had a hard job. Yet, it was important to him to have a job that was helpfull to animals.

1. _____   4. _____

2. _____   5. _____

3. _____   6. _____

Writing Activity

Write about a job in science you might want to have someday. Use at least three spelling words in your paragraph.

_____

_____

_____

_____

_____

_____

_____

Name_____

Look at the words in each set below. One word in each set
is spelled correctly. Look at Sample A. The letter next to the
correctly spelled word in Sample A has been shaded in. Do
Sample B yourself. Shade the letter of the word that is spelled
correctly. When you are sure you know what to do, go on with
the rest of the page.

Sample A:

- Ⓐ hopefull
- Ⓑ hopefill
- Ⓒ hopeful
- Ⓓ hopefle

Sample B:

- Ⓔ careles
- Ⓕ carelest
- Ⓖ carelless
- Ⓗ careless

1.
- Ⓐ careful
- Ⓑ carefull
- Ⓒ carefol
- Ⓓ carefill

2.
- Ⓔ cheerfil
- Ⓕ cheerful
- Ⓖ chearful
- Ⓗ cheerfull

3.
- Ⓐ helpfil
- Ⓑ helpfol
- Ⓒ helpful
- Ⓓ helpfull

4.
- Ⓔ colorfil
- Ⓕ colorfol
- Ⓖ colorfull
- Ⓗ colorful

5.
- Ⓐ harmfil
- Ⓑ harmfill
- Ⓒ harmfull
- Ⓓ harmful

6.
- Ⓔ peacfil
- Ⓕ peacefull
- Ⓖ peaceful
- Ⓗ peacefill

7.
- Ⓐ pitiful
- Ⓑ pitefull
- Ⓒ pitifill
- Ⓓ pitifull

8.
- Ⓔ painlass
- Ⓕ painless
- Ⓖ painliss
- Ⓗ painluss

9.
- Ⓐ pricelass
- Ⓑ pricelass
- Ⓒ priceluss
- Ⓓ priceless

10.
- Ⓔ helpless
- Ⓕ helpliss
- Ⓖ helplass
- Ⓗ helplest

11.
- Ⓐ sleeplass
- Ⓑ sleapless
- Ⓒ sleepless
- Ⓓ sleepliss

12.
- Ⓔ rainlass
- Ⓕ rainless
- Ⓖ raneless
- Ⓗ rainliss

13.
- Ⓐ helplessli
- Ⓑ helplissly
- Ⓒ helplesslly
- Ⓓ helplessly

14.
- Ⓔ carefulli
- Ⓕ carefuly
- Ⓖ carefully
- Ⓗ carefilly

15.
- Ⓐ peasefully
- Ⓑ peacefully
- Ⓒ peacefuly
- Ⓓ peacefilly

Name _____

Fold back the paper along the dotted line. Use the blanks to write each word as it is read aloud. When you finish the test, unfold the paper. Use the list at the right to correct any spelling mistakes.

1. _____
2. _____
3. _____
4. _____
5. _____
6. _____
7. _____
8. _____
9. _____
10. _____
11. _____
12. _____
13. _____
14. _____
15. _____

Review Words

16. _____
17. _____
18. _____

Challenge Words

19. _____
20. _____

1. because
2. rubber
3. about
4. puddle
5. alive
6. behind
7. before
8. around
9. better
10. attract
11. kettle
12. hammer
13. attend
14. tickle
15. people
16. peaceful
17. helpless
18. carefully
19. believe
20. beaver

At Home: Help your child practice the words he or she missed to prepare for the Posttest.

Name_____

Using the Word Study Steps

1. LOOK at the word.

2. SAY the word aloud.

3. STUDY the letters in the word.

4. WRITE the word.

5. CHECK the word.
   Did you spell the word right?
   If not, go back to step 1.

Find and Circle

Where are the spelling words?

X	C	Z	A	Q	T	E	R	A	W	P
A	B	O	U	T	Z	B	Z	T	P	U
B	E	C	A	U	S	E	B	T	H	D
Y	H	C	V	R	W	F	B	E	Z	D
R	I	V	F	A	R	O	U	N	D	L
U	N	H	K	W	Q	R	U	D	D	E
B	D	A	L	I	V	E	Y	Z	N	X
B	M	M	K	K	E	T	T	L	E	F
E	N	M	T	I	C	K	L	E	X	G
R	B	E	T	T	E	R	W	Q	T	Z
U	C	R	A	T	T	R	A	C	T	W
P	E	O	P	L	E	J	X	Z	J	W

 At Home: Review the Word Study Steps to help your child
   spell new words.

Name_____

because	puddle	before	attract	attend
rubber	alive	around	kettle	tickle
about	behind	better	hammer	people

Pattern Power!

This week's spelling words have syllables that are accented differently. Divide the spelling list into words where the first syllable is accented and where the second syllable is accented.

Accented first syllable

1. _____
2. _____
3. _____
4. _____
5. _____
6. _____
7. _____

Accented second syllable

8. _____
9. _____
10. _____
11. _____
12. _____
13. _____
14. _____
15. _____

Name_____

because	puddle	before	attract	attend
rubber	alive	around	kettle	tickle
about	behind	better	hammer	people

What's the Word?

Complete each sentence with a spelling word.

1. I am reading a book _____ butterflies.

2. _____ going to sleep, butterflies look for a safe place to rest.

3. It would _____ if a butterfly landed on your nose.

4. My mother bought a feeder that will _____ butterflies to our yard.

5. _____ have always found butterflies beautiful to watch.

6. Butterflies have eyes that allow them to see things that are _____ them.

7. The butterfly will drink from a mud _____.

8. If you are good, you can _____ the butterfly show with me.

9. I bought my mother a tea _____ with butterflies painted on it.

10. I love butterflies _____ they are so colorful.

11. It took us one hour to walk _____ the pond.

What Does It Mean?

Write a spelling word that matches each clue.

12. Strong, elastic material _____

13. A hand tool _____

14. Not dead _____

15. Something that is improved _____

Name_____

There are six spelling mistakes in this paragraph. Circle the misspelled words. Write the words correctly on the lines below.

Yesterday, I sat on the steps of school waiting for my mom to pick me up. No other peeple were arround. Most of the other kids were playing on the swing set. I didn't feel like playing. I was behind in science and my teacher was not happy. She said I had to do bedder on my science paper to get a good grade. The problem was I didn't know what to write abowt. Just then a butterfly flew up and landed on my hand. It was beautiful and had wings with orange and black stripes. It walked up my finger. The ticle made me laugh and it flew away. All of a sudden my mom drove up. I didn't get to show her the butterfly. But then I realized that I could write about butterflies for my paper! I asked my mom if we could stop at the library befour we went home. I couldn't wait to learn more about butterflies.

1. _____   4. _____

2. _____   5. _____

3. _____   6. _____

Writing Activity

Write about an experience you had with another creature. Use at least three spelling words in your paragraph.

_____

_____

_____

_____

_____

Name_____

Look at the words in each set below. One word in each set is spelled correctly. Look at Sample A. The letter next to the correctly spelled word in Sample A has been shaded in. Do Sample B yourself. Shade the letter of the word that is spelled correctly. When you are sure you know what to do, go on with the rest of the page.

Sample A:

Ⓐ beeside
Ⓑ beaside
Ⓒ beside
Ⓓ besighed

Sample B:

Ⓔ apple
Ⓕ apell
Ⓖ appel
Ⓗ appal

1. Ⓐ becawse
Ⓑ because
Ⓒ becoze
Ⓓ becauze

2. Ⓔ ruber
Ⓕ rubbir
Ⓖ rubber
Ⓗ rubbur

3. Ⓐ abowt
Ⓑ aboute
Ⓒ about
Ⓓ abbout

4. Ⓔ pudle
Ⓕ puddel
Ⓖ puddile
Ⓗ puddle

5. Ⓐ allive
Ⓑ alive
Ⓒ alaive
Ⓓ allave

6. Ⓔ behind
Ⓕ behund
Ⓖ behend
Ⓗ behinde

7. Ⓐ before
Ⓑ befure
Ⓒ befire
Ⓓ beffer

8. Ⓔ uhround
Ⓕ arond
Ⓖ arownd
Ⓗ around

9. Ⓐ beter
Ⓑ bettir
Ⓒ bettar
Ⓓ better

10. Ⓔ attract
Ⓕ atract
Ⓖ attrect
Ⓗ attrackt

11. Ⓐ kettle
Ⓑ kettal
Ⓒ kettel
Ⓓ ketel

12. Ⓔ hamer
Ⓕ hammur
Ⓖ hammir
Ⓗ hammer

13. Ⓐ atend
Ⓑ attend
Ⓒ attenned
Ⓓ atrand

14. Ⓔ tickel
Ⓕ tickul
Ⓖ tickle
Ⓗ tickal

15. Ⓐ people
Ⓑ poeple
Ⓒ peeple
Ⓓ peopel

© Macmillan/McGraw-Hill

Name _____

Read each sentence. If an underlined word is spelled wrong, fill in the circle that goes with that word. If no word is spelled wrong, fill in the circle below NONE. Read Sample A, and do Sample B.

NONE

A. In the <u>winter</u>, <u>peeple</u> walk <u>around</u> in heavy coats.
       A   B   C

A. Ⓐ ⬤Ⓑ Ⓒ Ⓓ

NONE

B. The <u>doctor</u> will check on my <u>sister</u> <u>laeter</u>.
      E   F   G

B. Ⓔ Ⓕ Ⓖ Ⓗ

NONE

1. My <u>sisster</u> likes it when there are many <u>people</u> <u>around</u>.
     A   B   C

1. Ⓐ Ⓑ Ⓒ Ⓓ

NONE

2. It is so <u>peaceful</u> in the <u>winter</u> when it is <u>abowt</u> to snow.
        E   F   G

2. Ⓔ Ⓕ Ⓖ Ⓗ

NONE

3. The <u>doctor</u> may <u>disagree</u> with the nurse about when you
     A   B

will be <u>able</u> to leave the hospital.
      C

3. Ⓐ Ⓑ Ⓒ Ⓓ

NONE

4. I am <u>unafraid</u> of starting <u>preskool</u> in the <u>winter</u>.
       E   F   G

4. Ⓔ Ⓕ Ⓖ Ⓗ

NONE

5. The <u>colorful</u> bird had <u>little</u> <u>perple</u> spots on his wings.
     A   B   C

5. Ⓐ Ⓑ Ⓒ Ⓓ

NONE

6. No one will be <u>abel</u> to see you if you stand <u>behind</u> the <u>cellar</u>
          E   F   G

door.

6. Ⓔ Ⓕ Ⓖ Ⓗ

NONE

7. She <u>carefully</u> walked <u>around</u> the <u>puttle</u> so she wouldn't get dirty.
     A   B   C

7. Ⓐ Ⓑ Ⓒ Ⓓ

NONE

8. <u>Peeple</u> <u>dislike</u> feeling that they are <u>helpless</u>.
   E   F   G

8. Ⓔ Ⓕ Ⓖ Ⓗ

NONE

9. The clerk gives a <u>nickel</u> for every <u>colurful</u> bottle you <u>return</u>.
              A   B   C

9. Ⓐ Ⓑ Ⓒ Ⓓ

NONE

10. What is <u>harmfull</u> <u>about</u> going to the <u>doctor</u>?
       E   F   G

10. Ⓔ Ⓕ Ⓖ Ⓗ

NONE

11. My <u>sister</u> hopes to <u>travul</u> to Europe <u>later</u> this year.
     A   B   C

11. Ⓐ Ⓑ Ⓒ Ⓓ

© Macmillan/McGraw-Hill

Name _____

12. He was <u>unifraid</u> of the dark <u>cellar</u> even when he was <u>little</u>.
            E   F   G

12. Ⓔ Ⓕ Ⓖ Ⓗ   NONE

13. It is <u>peaceful</u> to sit <u>behind</u> our <u>purple</u> house on the porch.
           A   B   C

13. Ⓐ Ⓑ Ⓒ Ⓓ   NONE

14. <u>Lader</u> in the <u>winter</u>, the <u>puddle</u> will turn to ice.
    E   F   G

14. Ⓔ Ⓕ Ⓖ Ⓗ   NONE

15. Little animals are <u>helples</u> in <u>harmful</u> weather.
     A   B   C

15. Ⓐ Ⓑ Ⓒ Ⓓ   NONE

16. I <u>dislike</u> it when mom has to leave me <u>behinde</u> to <u>travel</u>.
    E   F   G

16. Ⓔ Ⓕ Ⓖ Ⓗ   NONE

17. She must <u>carefuly</u> <u>return</u> the form to her <u>preschool</u> teacher.
          A   B   C

17. Ⓐ Ⓑ Ⓒ Ⓓ   NONE

18. I bet you a <u>nickel</u> that there is a <u>purple</u> monster in the <u>sellar</u>.
           E   F   G

18. Ⓔ Ⓕ Ⓖ Ⓗ   NONE

19. Most <u>people</u> <u>disagre</u> that children are totally <u>helpless</u>.
       A   B   C

19. Ⓐ Ⓑ Ⓒ Ⓓ   NONE

20. The <u>doctor</u> <u>carefully</u> wrapped a bandage <u>arownd</u> her leg.
      E   F   G

20. Ⓔ Ⓕ Ⓖ Ⓗ   NONE

21. My <u>sister</u> sold brownies for a <u>nickle</u> at the <u>preschool</u>
     A   B   C
bake sale.

21. Ⓐ Ⓑ Ⓒ Ⓓ   NONE

22. The <u>colorful</u> bird will <u>retern</u> to her nest <u>later</u>.
      E   F   G

22. Ⓔ Ⓕ Ⓖ Ⓗ   NONE

23. She is <u>unafraid</u> to <u>travel</u> in a calm and <u>peacefull</u> boat.
        A   B   C

23. Ⓐ Ⓑ Ⓒ Ⓓ   NONE

24. We <u>disagree</u> that you'll be <u>able</u> to hop over the <u>puddle</u>.
    E   F   G

24. Ⓔ Ⓕ Ⓖ Ⓗ   NONE

25. I <u>dislick</u> that she is <u>about</u> to go to that <u>harmful</u> place.
    A   B   C

25. Ⓐ Ⓑ Ⓒ Ⓓ   NONE

© Macmillan/McGraw-Hill